Reduction of Nitrogen Oxide Emissions

A C S   S Y M P O S I U M   S E R I E S **587**

Reduction of Nitrogen Oxide Emissions

Umit S. Ozkan, EDITOR
Ohio State University

Sanjay K. Agarwal, EDITOR
Exxon Chemical Company

George Marcelin, EDITOR
Altamira Instruments

Developed from a symposium sponsored
by the Division of Petroleum Chemistry, Inc.,
at the 207th National Meeting
of the American Chemical Society,
San Diego, California,
March 13–17, 1994

American Chemical Society, Washington, DC 1995

Library of Congress Cataloging-in-Publication Data

Reduction of nitrogen oxide emissons / Umit S. Ozkan, editor, Sanjay K. Agarwal, editor, George Marcelin, editor.

p.   cm.—(ACS symposium series, ISSN 0097–6156; 587)

"Developed from a symposium sponsored by the Division of Petroleum Chemistry, Inc., at the 207th National Meeting of the American Chemical Society, San Diego, California, March 13–17, 1994."

Includes bibliographical references and index.

ISBN 0–8412–3150–8

1. Nitrogen oxides—Environmental aspects—Congresses. 2. Air quality management—Congresses.

I. Ozkan, Umit S., 1954–   . II. Agarwal, Sanjay K., 1965–   . III. Marcelin, George, 1948–   . IV. American Chemical Society. Division of Petroleum Chemistry. V. American Chemical Society. Meeting (207th: 1994: San Diego, Calif.) VI. Series.

TD885.5.N5R43   1995
628.5'32—dc20   95–2400
                                                                           CIP

This book is printed on acid-free, recycled paper.

Foreword

THE ACS SYMPOSIUM SERIES was first published in 1974 to provide a mechanism for publishing symposia quickly in book form. The purpose of this series is to publish comprehensive books developed from symposia, which are usually "snapshots in time" of the current research being done on a topic, plus some review material on the topic. For this reason, it is necessary that the papers be published as quickly as possible.

Before a symposium-based book is put under contract, the proposed table of contents is reviewed for appropriateness to the topic and for comprehensiveness of the collection. Some papers are excluded at this point, and others are added to round out the scope of the volume. In addition, a draft of each paper is peer-reviewed prior to final acceptance or rejection. This anonymous review process is supervised by the organizer(s) of the symposium, who become the editor(s) of the book. The authors then revise their papers according to the recommendations of both the reviewers and the editors, prepare camera-ready copy, and submit the final papers to the editors, who check that all necessary revisions have been made.

As a rule, only original research papers and original review papers are included in the volumes. Verbatim reproductions of previously published papers are not accepted.

M. Joan Comstock
Series Editor

Contents

Preface

Reduction of nitrogen oxide emissions from stationary and mobile sources continues to pose technological and scientific challenges. The public is widely aware of problems such as ozone depletion in the stratosphere, the greenhouse effect, and air pollution. Concern for the environment coupled with stricter NO_x emission standards adopted by many countries lends a renewed urgency to these challenges. While work on existing technologies continues to provide a better insight into the fundamental phenomena involved, recent research is opening up new possibilities for reducing NO_x emissions. The next decade very likely will witness the emergence of new technologies based on new fundamental research.

The symposium that is the basis of this book provided a forum where a broad spectrum of studies tackling the NO_x reduction problem were presented and discussed. Following the format of the symposium, this book brings together different approaches to the control of nitrogen oxide emissions. We hope that this book will provide a quick overview of some of the work already in progress and will stimulate further research in the area. We believe that it may interest a broad audience, including catalysis researchers, environmental policy makers, plant managers, and pollution control specialists.

Some of the chapters provide new insights to the previously studied NO_x reduction techniques. Others point to new and exciting possibilities for future directions. In addition to the two overview chapters that focus on the technical and the regulatory aspects of nitrogen oxide emission control, the book contains reports on selective catalytic reduction of NO with NH_3; use of CO, hydrogen, methane, and other hydrocarbons as reducing agents; and catalytic direct decomposition of NO. The catalytic systems studied include aerogels of titania, silica, and vanadia; zeolites; Ce-, Cu-, and Na-exchanged ZSM-5; Fe silicate; silica-supported Ce and Rh; and Cu catalysts coprecipitated with ZrO_2 and Ga_2O_3. Most of the chapters deal with catalytic solutions, but two interesting examples of noncatalytic techniques for the reduction of the nitrogen oxide emissions are also presented.

We express our gratitude to all the authors who worked with us patiently in meeting the deadlines to complete the book. We are grateful to the Division of Petroleum Chemistry, Inc., of the American Chemical

Society for sponsoring the symposium on which this book was based. We also thank Anne Wilson of the ACS Books Department for providing encouragement and assistance at every stage of preparing the book.

UMIT S. OZKAN
Department of Chemical Engineering
The Ohio State University
Columbus, OH 43210

SANJAY K. AGARWAL
Exxon Chemical Company
Box 536
Linden, NJ 07036

GEORGE MARCELIN
Altamira Instruments
2090 William Pitt Way
Pittsburgh, PA 15238

December 8, 1994

Chapter 1

NO$_x$ Control from Stationary Sources
Overview of Regulations, Technology, and Research Frontiers

Carmo J. Pereira and Michael D. Amiridis

W. R. Grace & Co.–Connecticut, Research Division, 7379 Route 32, Columbia, MD 21044

Environmental regulations are requiring the commercial deployment of technologies for the control of nitrogen oxides emissions from stationary sources and are creating opportunities for the development of new technologies. This chapter provides an industrial perspective on the regulatory picture, discusses commercially-available technologies, and identifies some future research opportunities.

The 1992 U.S. emissions of anthropogenic nitrogen oxides (NOx) are estimated at 23 million tons. Of this amount, approximately 45% were from transportation sources (cars, trucks, etc.) and the remainder from stationary sources. Examples of stationary source emitters include power plants (53%), internal combustion engines (20%), industrial boilers (14%), process heaters (5%), and gas turbines (2%). Total NOx emissions are estimated to have increased 5% since 1983. Stationary sources have accounted for the majority of the increase; emissions from mobile sources have remained relatively constant. Approximately 51% of the total NOx emissions are a result of combustion in stationary-sources applications [1].

Even though nitrogen appears in a wide variety of oxidation states, ranging from +5 (in HNO$_3$) to -3 (in NH$_3$), with intermediate oxidation states of +4 (NO$_2$), +2 (NO), +1 (N$_2$O), and 0 (N$_2$), nitrogen oxides emissions subject to regulations are nitric oxide (NO) and nitrogen dioxide (NO$_2$), together referred to as NOx. NOx is formed by oxidation of nitrogen-containing compounds in the fuel (fuel NOx), by reaction of air-derived nitrogen and oxygen (thermal NOx) and by reaction between radicals in the combustion flame (prompt NOx). Under high temperature combustion conditions, thermodynamics favor NO formation; consequently, less than 10% of the NOx in typical exhausts is in the form of NO$_2$. NO, however, converts to NO$_2$ in the atmosphere. Though not an important constituent in combustion exhausts, N$_2$O has also received attention for troposcopic ozone destruction and as a greenhouse gas; however, N$_2$O emissions are not currently regulated. The identification and subsequent control of sources of NOx are becoming increasingly important in the U.S.

0097–6156/95/0587–0001$12.00/0
© 1995 American Chemical Society

Examples of stationary sources that generate NOx emissions include: rich-burn, stoichiometric or lean-burn engines (e.g., for compressing pipeline natural gas), utility boilers (e.g., for power generation), gas turbine cogeneration plants (e.g., for producing process steam and electricity), process heaters (e.g., for operating refinery reactors) and chemical plants (e.g., that make nitric acid). The convention used for reporting NOx emissions varies depending upon the application: pounds per brake horsepower-hour for engines, ppmvd @ 3% O$_2$ for utility boilers, ppmvd @ 15% O$_2$ for cogeneration plants, pounds per million BTU for process heaters, and tons per year for point-source emissions.

The composition of the exhaust depends on the type and composition of the fuel and on combustion conditions. The main components are O$_2$ (0-15%), CO$_2$ (3-12%), H$_2$O (6-18%) and N$_2$. Typical ranges of air pollutants from the combustion of natural gas, oil and coal are shown in Table I. In addition to NOx, commonly encountered pollutants include carbon monoxide (CO), hydrocarbons or Volatile Organic Compounds (VOCs), sulfur oxides (SOx) and particulates. Municipal Solid Waste (MSW) incinerator or waste-to-energy plant exhausts may also contain acid gases (e.g., HCl, HF), dioxins, furans and trace amounts of toxic metals such as mercury, cadmium and lead.

Table I. Ranges of Air Pollutants in Exhaust Gases

	Gas	Oil	Coal
CO, ppm	10-100	10-200	<50
HCs, ppm	1-10	1-10	1-10
NOx, ppm	25-150	60-1,000	200-1,000
SOx, ppm	<0.5	5-500	500-2,000
Particulate, gr/dscf	NA	0.01-0.1	2-10

Driving forces for controlling NOx emissions include:

a. Health effects: NO$_2$ can irritate the lungs and lower resistance to respiratory infection (such as influenza). Frequent exposure to concentrations higher than those normally found in the ambient air may cause increased incidence of acute respiratory disease in children.

b. Acid precipitation: NOx and sulfur oxides (SOx) emissions are primary contributors to acid rain, which is associated with a number of effects including acidification of lakes and streams, accelerated corrosion of buildings and monuments, and visibility impairment.

c. Atmospheric ozone non-attainment: NOx and VOCs react in the atmosphere to form ozone, a photochemical oxidant and a major component of smog. These reactions are accelerated by sunlight and temperature; therefore, peak ozone levels occur in the summer. Atmospheric ozone can cause respiratory problems by damaging lung tissue, reducing lung function and sensitizing the lungs to other irritants. The decrease in lung function is often accompanied by symptoms such as chest pain, coughing, nausea and pulmonary congestion.

Ozone is also responsible for agricultural crop yield loss and can cause noticeable foliar damage in crops and trees.

In this chapter, we provide an industrial perspective on the complex regulatory picture, discuss commercially-available technologies for NOx control from stationary sources, and identify some future research opportunities.

Regulations

The Clean Air Act Amendments of 1970, which followed the original Clean Air Act of 1967, set national air quality standards for six criteria air pollutants: NOx, SOx, ozone, carbon monoxide (CO), particulates and lead. The result was the removal of lead from gasoline and the installation of emission control technologies, including baghouse filters for particulate control, wet and dry scrubbers for SOx control and automobile exhaust catalysts for controlling hydrocarbons (HC), CO and NOx. As a consequence, lead emissions have been dramatically reduced, SOx emissions are being controlled, and automobile CO, HC and NOx emissions have decreased by nearly a factor of 10 (over uncontrolled emissions). In spite of these dramatic improvements, in 1989 approximately 130 million people in the U.S. lived in 96 areas which did not meet air quality standards either in ozone, in carbon monoxide, or in both [2].

The 1990 Clean Air Act Amendments (CAAA) were enacted to address various air pollution concerns: non-attainment for criteria pollutants (Title I), mobile source emissions (Title II), air toxic emissions (Title III), and acid rain (Title IV). The 1990 CAAA's NOx reduction target is 2 million tons/year over 1980 levels by the year 2000. The Ammendments overlay a complex existing regulatory picture. The specific NOx standard a site must meet depends on a variety of different factors including application, fuel, site status (e.g. new or retrofit), and the location. For example, new sources must often meet more stringent New Source Performance Standards (NSPS). State Implementation Plans (SIP) aimed at alleviating local non-attainment problems, can vary from state to state and may call for more stringent controls than required by the federal regulations.

Atmospheric ozone is produced as a result of complex reactions involving NOx and VOCs [3]. The concentration of ozone is a non-linear function of the concentrations of these two reactants. This relationship is demonstrated in the typical ozone isopleth diagram shown in Figure 1. The diagram exhibits a diagonal ridge from the lower left to the upper right corner of the graph that corresponds to a VOC/NOx ratio of approximately 8. As a result, the optimal strategy for reducing ambient ozone is related to the local concentrations of NOx and VOCs. For VOC/NOx > 8 the optimal strategy for controlling ozone is to control NOx. Alternatively, when the VOC/NOx ratio is < 8, the best approach is to control VOCs. There has been continuing debate about total VOC emission levels. A recent National Research Council study [3] has concluded that anthropogenic VOC emissions are likely to have been underestimated by a factor of 2 to 4. A possible consequence of this finding could be the need for more stringent NOx control regulations.

Typical NOx reductions required can vary between 30 and 90%. In ozone non-attainment areas, NOx emission reductions of greater than 80% can be required. The South Coast Air Quality Management District (SCAQMD) in Southern California has regulations that are among the most stringent in the world. Similar local regulations

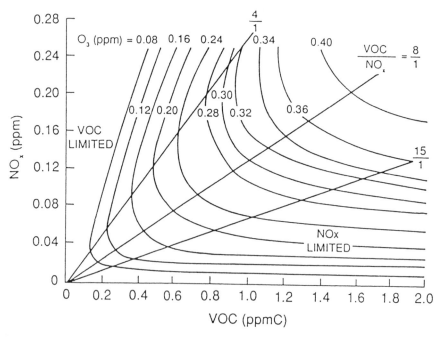

Figure 1: Typical ozone isopleths [from ref. 3].

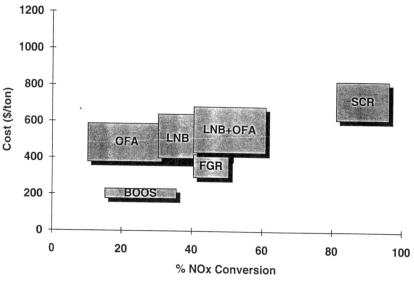

Figure 2: Cost of NOx control for 500 MW base load oil and gas-fired utility boilers (BOOS: Burners-Out-Of-Service, OFA: Over-Fired Air, LNB: Low NOx Burners, FGR: Flue Gas Recirculation, SCR: Selective Catalytic Reduction) [adapted from ref. 7].

have also been proposed by a group of northeastern states referred to as the North Eastern States for Coordinated Air Use Management (NESCAUM). An example of these regulations is shown in Table II.

Table II. Gas Turbine NOx Emissions Limits @ 15% Oxygen
[adapted from refs. 4,5].

A. SCAQMD Rule 1134	
Unit Size (MW)	NOx Limit (ppm)
0.3-2.9	25
2.9 -10	9
> 10	9
> 60 (combined cycle)	9
B. NESCAUM Recommendations	
Rating (MMBtu/hr)[a]	NOx Limit (ppm)
1 - 100	42
> 100	9

[a]100 MMBtu/hr corresponds to a 10 MW unit.

The regulations factor in the economic considerations associated with installation of available control technologies. Small facilities are typically excluded from the regulations or allowed to have higher emission limits [6]. Typically, the cost per ton of NOx reduced increases with percent NOx conversion. Some recent information on the cost effectiveness of several commercial technologies used in gas-fired utility boiler service is shown in Figure 2. A common regulatory approach is to specify compliance limits and then let the operator select the appropriate control technology. As a result, there is an incentive for developing new or improved technologies that achieve high emissions reductions at low cost.

Competitive pressures and improved technology are lowering the cost of control equipment (e.g., see [8]) and stimulating the development of new technologies. A creative approach for improving air quality is to let the marketplace determine the value of NOx emissions [9]. This is accomplished by the establishment of NOx credits. Depending on the market value of a credit, it may be to an operator's advantage to reduce emissions beyond the regulatory requirements and to sell credits to another operator that is not in compliance.

The true impact of the 1990 CAAA is only now beginning to be felt: many of the regulations do not kick in fully until 1995 or beyond. Nevertheless, recent emissions and air quality data are encouraging. For the first time in 1992 no areas in the U.S. had measured values exceeding the NOx ambient standard [1]. Further improvements in air quality are likely as fuel reformulation regulations and other mobile source programs take effect.

Table III. Examples of Vanadia/titania Catalyst Technology Development

Goal	Strategy	Issues	Result
Maximize NOx conversion	Increase intrinsic activity Optimize catalyst composition (vanadia loading, activity promoters)	Activity for SO$_2$ oxidation Waste disposal	V$_2$O$_5$-WO$_3$/TiO$_2$ and V$_2$O$_5$-MoO$_3$/TiO$_2$ catalysts
	Increase intrapellet mass transfer Optimize macroporosity	Mechanical properties (crush strength, abrasion resistance)	Bimodal TiO$_2$ and TiO$_2$-SiO$_2$ supports
	Increase catalyst residence time Increase cell density	Plugging by particulates	Optimized cell densities for different applications
Decrease SO$_2$ oxidation	Optimize catalyst composition (vanadia loading, SO$_2$ activity suppresents)	Activity for NOx reduction Waste disposal	Low V$_2$O$_5$ loading catalysts
Decrease NH$_3$ slip	Improve NH$_3$ distribution system Increase catalyst volume	Cost	NH$_3$ injection systems design Establish relationship between activity, space velocity, NH$_3$/NOx ratio and NH$_3$ slip.
Improve catalyst durability	"Protective" additives Bimodal, high surface area supports	Impact on NOx activity and SO$_2$ selectivity	V$_2$O$_5$-MoO$_3$/TiO$_2$ catalysts High surface area, bimodal supports
	Strength and abrasion resistance enhancers		Additives and face hardening methods
Minimize pressure drop	Catalyst and reactor design	High NOx activity Cost	Thin-walled monolith catalysts Reactor designs that increase catalyst cross-sectional area in duct

Technology

NOx emissions may be controlled by primary or secondary measures. Primary measures are aimed at reducing the formation of NOx. Examples of primary measures include fuel switching (e.g., moving from coal to oil or to gas) and in-combustion modifications. Examples of in-combustion modification such as Low NOx Burners (LNB), Over-Fired Air (OFA), Burners-Out-Of-Service (BOOS), Flue Gas Recirculation (FGR), etc. have been reviewed in the literature, e.g., see [10]. Secondary measures reduce NOx after it is formed. An example, Selective Non-Catalytic Reduction (SNCR), reduces NOx via ammonia injection at temperatures of between 1500 and 1700°F [11]. The use of reductants other than ammonia, such as urea [12] and cyanuric acid [13], has also been discussed.

Selective Catalytic Reduction (SCR) using ammonia as the reductant provides NOx reduction levels of greater than 80%. Three types of catalyst systems have been deployed commercially: noble metal, base metal and zeolites. Noble metals are typically washcoated on inert ceramic or metal monoliths and used for particulate-free, low sulfur exhausts. They function at the lower end of the SCR temperature range (460-520°F) and are susceptible to inhibition by SOx [14]. Base metal vanadia-titania catalysts may either be washcoated or extruded into honeycombs [11]. Typically washcoated catalysts are only used for treating particulate-free, clean gas exhausts. Extruded monoliths are used in particulate-laden coal and oil-fired applications. The temperature window for these catalysts is 600-750°F. Zeolites may also be washcoated or extruded into honeycombs. They function at relatively high temperatures of 650-940°F [15]. Zeolites may be loaded with metal cations (such as Fe, Cu) to broaden the temperature window [16].

The most widely-used commercial SCR catalyst is based on vanadia supported on high surface area anatase titania. Base metal oxides (e.g., WO_3) may be added to the catalyst to increase activity and widen the temperature window, to increase poison resistance and to reduce ammonia as well as SO_2 oxidation [17]. Conventional titania-based SCR catalyst performance has been found to be limited by the diffusion of reactants into the walls of the monolith. An advance in extruded SCR catalyst technology is the development of titania-silica-based catalysts with higher macroporosity which have a higher volumetric activity than the titania-based catalysts [18]. Both the SCR catalyst and the process have been optimized over the last two decades. The key design considerations include catalytic performance, durability and physical requirements (Table III). A concern for deployment of SCR catalysts near urban areas is the transportation, storage and handling of anhydrous ammonia. The recent trend encourages the use of aqueous ammonia [19].

In the case of utility boilers, the catalyst accounts for less than 50% of the overall capital cost of the SCR system [20]. The percentage of non-catalyst costs can be even higher in retrofit applications in which existing equipment has to be moved to accommodate the SCR system. Approaches being considered to minimize system costs in these applications include: a) creative reactor designs for placing the catalyst in the existing boiler train (e.g., in one case, the catalyst was placed at a 60° angle to the horizontal to increase cross-sectional area [21]), and b) hybrid systems in which the overall NOx reduction is achieved through a combination of technologies. An

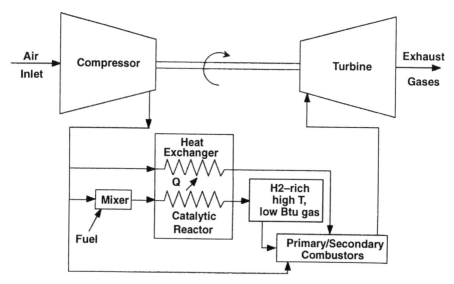

Figure 3: Schematic of hybrid catalytic combustion [from ref. 28].

example is the demonstration project at San Diego Gas & Electric's Encina plant that uses urea-based SNCR, in-duct SCR catalyst and catalyzed Ljungstrom air preheater elements [10].

Research Frontiers

The most obvious approach for NOx control is to minimize its formation. There is considerable research activity on NOx reduction via combustion modification. One such approach, is the low NOx burner developed by Beer and coworkers, in which gradual mixing of burner air with the centrally injected fuel is achieved by radial stratification of the flame. This stratification is brought about by a combination of swirling air flow and strong positive radial density gradients in the flame. The NOx emissions (@3% O_2) of 70 ppm without FGR and 15 ppm with 32% FGR have been achieved for a 1 MW-size burner [22].

Catalytic combustion was originally discovered in the early 1970's [23]. The concept is based on the observation that catalysts can be used to light off combustion reactions and to sustain gas-phase oxidation reactions after the catalyst. Peak temperatures obtained via catalytic combustion are significantly lower than those achieved in admixed or premixed flames; consequently, NOx emissions are lower. Catalyst requirements include low light-off temperature, low reactor backpressure at high volumetric flow rates, high temperature stability, thermal shock resistance and durability.

Recent advances in catalytic combustion include the development of new materials having high thermal stability and novel catalytic reactors [24]. Currently proposed reactor designs utilize two or more types of oxidation catalysts in series, with each catalyst fulfilling a specific purpose (e.g., the 1st-stage catalyst is tailored for low temperature methane light-off, the 2nd-stage catalyst is engineered to control methane conversion, the 3rd-stage catalyst has high thermal stability, etc.) [e.g., 25]. Hybrid heat exchanger/catalyst reactors having only a fraction of the metal monolith structure coated with catalyst have been developed to control the maximum temperature reached within the device [26,27].

A relatively new concept, referred to as "hybrid catalytic combustion", that combines innovative reactor design with air staging is also being investigated. The concept is illustrated in Figure 3. Natural gas fuel and 7.5% of the combustion air are premixed and sent to a catalytic reactor/heat exchanger device. The fuel-rich mixture flows through the partial oxidation reactor which is cooled by 50% of the combustion air. The reaction products together with the remaining (42.5%) air are further oxidized in a primary zone, still under fuel-rich conditions. Oxidation products are finally mixed with the (50%) heat exchanged air, combusted in a secondary zone, and then sent to a gas turbine. NOx emissions are reduced because: (1) lower peak temperatures reduce thermal NOx, and (2) catalytic partial oxidation to CO and H_2 reduces prompt NOx [28].

The preferred method for controlling NOx, once formed, is to catalytically decompose it to nitrogen and oxygen. Although thermodynamically favored, steady-state direct NO decomposition in the presence of excess O_2 has only been demonstrated recently. Iwamoto and co-workers have demonstrated that Cu-exchanged zeolites, and in particular Cu-ZSM5, are the most active catalysts for this reaction [29]. The NO decomposition activity of Cu-ZSM5 has been attributed to the high reducibility of the Cu cations in the ZSM5 environment that allows the catalyst to continuously desorb oxygen [30]. Unfortunately, NO decomposition activity of Cu-ZSM5 is inhibited by O$_2$ and H$_2$O and poisoned by SO$_2$. The development of NO decomposition catalysts that function in typical exhaust environments still represents a research challenge.

Another recent finding is that NOx can be selectively reduced by a variety of organic compounds (e.g., alkanes, olefins, alcohols) over several different catalysts under excess O$_2$ ("lean") conditions. This reaction is referred to as hydrocarbon-SCR (HC-SCR). HC-SCR was first independently reported by Held [31] and by Iwamoto [32] and their co-workers. Both groups concluded that Cu-ZSM5 was the most effective catalyst for this reaction. Unlike NO decomposition, however, the Cu-ZSM5 HC-SCR catalyst is not completely poisoned by SO$_2$[33]. Follow-up studies have now demonstrated that various other hydrogen- [34] and metal-exchanged zeolites [35-37], as well as non-zeolitic materials [38-40] are HC-SCR catalysts under a variety of reaction conditions.

Considerable research work is currently ongoing on the development of HC-SCR catalysts. Cu-ZSM5 has attracted the most attention in the literature. A significant technical hurdle for this catalyst is that it deactivates under hydrothermal aging conditions as a result of zeolite dealumination [41,42]. Methods for stabilizing the zeolite are being investigated. In addition to durability issues, Cu-ZSM5 has been reported to form HCN when ethylene and propylene are used as reductants [43]. Other more obvious selectivity issues will also have to be addressed: the partial oxidation product CO is a criteria pollutant and unburnt hydrocarbon slip will likely be regulated under the CAAA statutes. Work aimed at elucidating the HC-SCR reaction mechanism could provide useful leads for developing catalysts having high activity, a broad operating temperature window, and acceptable durability. Such catalysts, once developed, will likely be used for both mobile and stationary NOx emission control. In stationary emission control the nearer-term use will probably be for lean-burn stationary engines, while the longer-term use will be for utility boilers and cogeneration plants.

Some of the hydrothermal stability and the partial oxidation issues of Cu-ZSM5 appear to be resolved by the use of noble metal-based HC-SCR catalysts [44-46]. A research challenge is to reduce the formation of N$_2$O, in favor of N$_2$. While active at much lower temperatures, these catalysts have been shown to be sensitive to inhibition by SO$_2$ in NH$_3$-SCR [14].

In addition to the higher molecular weight hydrocarbons, it has been further reported that at least three zeolites, Co- [47], Ga- [48], and In-ZSM5 [49], can activate methane (CH$_4$) and utilize it as a selective NOx reducing agent under excess O$_2$ conditions. The turnover frequencies of these catalysts are estimated to be lower than the corresponding turnover frequencies for the reduction of NOx by olefins, which in turn are approximately one order of magnitude lower than the turnover frequencies of vanadia/titania NH$_3$-SCR catalysts. Nevertheless, the alternative of utilizing CH$_4$ as a reducing agent instead of NH$_3$ is extremely attractive: (a) it is relatively inexpensive and readily available, and (b) CH$_4$ emissions are currently not regulated. Some of the economic and technical hurdles in developing HC-SCR (or CH$_4$-SCR) catalysts will be similar to those encountered in the development of vanadia-titania catalysts (e.g., see Table III). Cost-effective methane SCR catalysts, if developed, will likely replace ammonia SCR catalysts for NOx control.

Coal combustion is a significant source of NOx emissions and the largest source of SOx emissions as well. Thus, a discussion of NOx control technologies would be incomplete without reference to technologies being developed for simultaneous NOx and SOx control. Examples of such technologies include wet scrubbing, modified spray-dryer scrubbing, in-duct sorbent injection, adsorption, e-beam, and activated coke processes. A detailed discussion on the status of these technologies is beyond the scope of this chapter and can be found in a recent article by Livengood and Markussen [50].

Conclusions

Tightening environmental regulations are requiring the deployment of cost-effective NOx control technologies and are providing opportunities for the development of new technologies. Promising research horizons include low-NOx burners, catalytic combustion, NO decomposition, hydrocarbon (or methane)-SCR and simultaneous NOx and SOx control .

Acknowledgments

C. J. Pereira would like to acknowledge useful discussions with Dr. L. L. Hegedus during the preparation of an earlier manuscript. The authors would like to thank Dr. R. H. Harding, Dr. J. E. Kubsh, M. Uberoi and Mr. K. P. Zak all of W. R. Grace & Co.-Conn. for their helpful comments.

Literature Cited

1.   "National Air Quality and Emissions Trends Report, 1992", U.S. Environmental Protection Agency, Research Triangle Park, North Carolina, October 1993.
2.   "National Air Quality and Emissions Trends Report, 1989", U.S. Environmental Protection Agency, Research Triangle Park, North Carolina, February 1991.
3.   "Rethinking the Ozone Problem in Urban and Regional Air Pollution", National Research Council, National Academy Press, Washington, DC, 1991.
4.   "Emission of Oxides of Nitrogen from Stationary Gas Turbines", South Coast Air Quality Management District, Rule 1134, August 1989.
5.   "Recommendation on Emission Limits for Gas Turbines", Northeast States for Coordinated Air Use Management, October 1988.
6.   Swingle, R.L., Gavelick, B., Proc. Council of Industrial Boiler Owners (CIBO) NOx Control VII and Permitting Conference, Oakbrook, Illinois, 1994.
7.   "NOx Control Options Guidebook", STAPPA/ALAPCO, June 1994.
8.   Gouker, T.R., Solar, J.P., Brundrett, C.P., EPA/EPRI Joint NOx Control Symposium, San Fransisco, California, 1989.
9.   Leone, M., *Power*, 9, December 1990.
10. Makansi, J., *Power*, 11, May 1993.
11. Bosch, H., Janssen, F., *Catalysis Today*, **2**, 369 (1987).
12. Abele, A.R., Kwan, Y., Mansour, M.N., Kertamus, N.J., Radak, L.J., Nylander, J.H., 1991 Joint Symposium on Stationary Combustion NOx Control, Washington, DC,   1991.
13. Wicke, B.G., Grady, K.A., Ratcliffe, J.W., *Nature*, **338**, 492 (1989).
14. Pereira, C. J., Gulian, F. J., Czarnecki, L. J., Rieck, J.S., Proc. AWMA Annual

Meeting, Paper No. 90-105.6, Pittsburgh, Pennsylvania, June 1990.

15. Kiovsky, J.R., Koradia, P.B., Lim, C.T., *Ind. Eng. Chem. Prod. Res. Dev.*, **19**, 218 (1980).
16. Byrne, J.W., Chen, J.M., Speronello, B.K., *Catalysis Today*, **13**, 33 (1992).
17. Chen, J.P., Yang, R.T., *Appl. Catal. A: General*, **80**, 135 (1992).
18. Beeckman, J.W., Hegedus, L.L., *Ind. Eng. Chem. Res.*, **30**, 969 (1991).
19. Fogman, C.B., Brummer, T.A., *Hydrocarbon Processing*, 120, August 1991.
20. Boer, F.P., Hegedus, L.L., Gouker, T.R., Zak, K.P., *CHEMTECH*, **20**, 312 (1990).
21. Johnson, L., Negrea, S., Ghoreishi, F., Proc. EPRI/EPA 1993 Joint Symposium on Stationary Combustion NOx Control, Miami Beach, Florida, May 1993.
22. Toqan, M.A., Beer, J.M., Jansohn, P., Sun, N., Testa, A., Shihadeh, A., Teare, J.P., Proc. Twenty Fourth Symposium (International) on Combustion, The Combustion Institute, p. 1391, (1992).
23. Pfefferle, W.C., U.S. Patent 3,928,961, assigned to Engelhard Minerals & Chemicals, 1975.
24. Zwinkels, M.F.M., Jaras, S.G., Menon, P.G., Griffin, T.A., *Catal. Rev. - Sci. Eng.*, **35**,319 (1993).
25. Dalla Beta, R.A., Ezawa, N., Tsurumi, K., Schlatter, J.C., Nickolas, S.G., U.S. Patent 5,183,401, assigned to Catalytica, Inc. and Tanaka Kikinzoku Kogyo KK, 1993.
26. Retallick, W.B., Alcorn, W.R., U.S. Patent 5,202,303, assigned to W.R. Grace & Co.-Conn., 1993.
27. Dalla Beta, R.A., Ribeiro, F.W., Tsurumi, K., Ezawa, N., Nickolas, S.G., U.S. Patent 5,250,489, assigned to Catalytica, Inc. and Tanaka Kikinzoku Kogyo KK, 1993.
28. Colket, M.B., Kesten, A.S., Sangiovanni, J.J., Zabielski, M.F., Pandy, D.R., Seery, D.J., U.S. Patent 5,235,804, assigned to United Technologies Corporation, 1993.
29. Iwamoto, M., Proc. International Symposium on Chemistry of Microporous Crystals, Tokyo, June 1990.
30. Li, Y., Hall, W.K., *J. Catal.*, **129**, 202 (1991).
31. Held, W., Konig, A., Richter, T., Puppe, L., *SAE Technical Paper Series*, SAE Paper 900496 (1990).
32. Iwamoto, M., Proc. Meeting on Catalytic Technology for Removal of Nitrogen Monoxide, p.17, Tokyo, January 1990.
33. Iwamoto, M., Yahiro, H., Shundo, S., Yu-u, Y., Mizuno, N., *Appl. Catal.*, **69**, L15 (1991).
34. Hamada, H., Kintaichi, Y., Sasaki, M., Ito, T., Tabata, M., *Appl. Catal.*, **64**, L1 (1990).
35. Misono, M., Kondo, K., *Chem. Lett.*, 1001 (1991).
36. Kikuchi, E., Yogo, K., Tanaka, S., Abe, M., *Chem. Lett.*, 1063 (1991).
37. Sato, S., Hirabayashi, H., Yahiro, H., Mizuno, N., Iwamoto, M., *Catal. Lett.*, **12**, 193 (1992).
38. Nakatsuji, T., Shimizu, H., Yasukawa, R., Proc. 1st Japan-EC Joint Workshop on the "Frontiers of Catalytic Science and Technology", p. 274, Tokyo (1991).
39. Truex, T.J., Searles, R.A., Sun, D.C., *Platinum Metals Rev.*, **36**, 2 (1992).
40. Kung, M.C., *Catal. Lett.*, **18**, 111 (1993).
41. Monroe, D.R., DiMaggio, C.L., Beck, D.D., Matekunas, F.A., *SAE Technical Paper Series*, SAE Paper 930737 (1993).
42. Grinsted, R.A., Jen, H.-W., Montreuil, C.N., Rokosz, M.J., Shelef, M., *Zeolites*, **13**, 602 (1993).
43. Radtke, F., Koeppel, R.A., Baiker, A., *Appl. Catal. A: General*, **107**, L125 (1994).

44. Hirabayashi, H., Yahiro, H., Mizuno, N., Iwamoto, M., *Chem. Lett.*, 2235 (1992).
45. Engler, B.H., Leyrer, J., Lox, E.S., Ostgathe, K., *SAE Technical Paper Series*, SAE Paper 930735 (1993).
46. Ansell, G.P., Golunski, S.E., Hayes, J.W., Walker, A.P., Burch, R., Millington, P.J., Preprints Third International Congress on Catalysis and Automotive Pollution Control, Vol. 1, p.255, Belgium (1994).
47. Li, Y., Armor, J.N., *Appl. Catal. B: Environ.*, **1**, L31 (1992).
48. Kikuchi, E., Yogo, K., Proc. International Forum on Environmental Catalysis, Tokyo 1993.
49. Yogo, K., Ono, T., Terasaki, I., Egashira, M., Okazaki, N., Kikuchi, E., *Shokubai*, **36**, 95 (1994).
50. Livengood, C.D., Markuson, J.M., Proc. of the 1993 A&WMA/EPRI/ASME International Symposium, New Orleans, Louisiana, March 1993.

RECEIVED November 14, 1994

Chapter 2

Clean Air Act Requirements: Effect on Emissions of NO$_x$ from Stationary Sources

Doug Grano

Office of Air Quality Planning and Standards, U.S. Environmental Protection Agency (MD-15), Research Triangle Park, NC 27711

The 1990 Clean Air Act (CAA) includes substantial new requirements to reduce emissions of nitrogen oxides (NO$_x$) from major stationary sources. To help attain the ozone air quality standard in the near-term, certain existing sources must install reasonably available control technology (RACT) and new sources must install controls representing the lowest achievable emission rate. To reduce acid rain, coal-fired utility boilers must meet emission limits in two phases. In a longer-term, the CAA requires States to adopt additional control measures as needed to attain the ozone standard. Many of these attainment strategies will include new NO$_x$ controls which achieve emission reductions much greater than the NO$_x$ RACT/phase I acid rain limits. The CAA NO$_x$ requirements are new and experience with NO$_x$ controls is limited. There is a clear need for the development of NO$_x$ control systems to meet near- and long-term CAA requirements.

The Clean Air Act (CAA) as amended in 1990 includes substantial new requirements to reduce emissions of nitrogen oxides (NO$_x$) from major stationary sources. The CAA is organized into several titles. Titles I and IV are discussed in this paper since they regulate NO$_x$ emissions from major stationary sources. This paper summarizes existing EPA guidance and regulatory requirements and references the relevant EPA Federal Register publications, documents, and memoranda. As described below, Title I is directed at attaining and maintaining the National Ambient Air Quality Standards (NAAQS) and Title IV's primary purpose is to reduce acid deposition.

Purposes of Titles I and IV of the CAA

The primary purpose of NO_x controls under Title I is to reduce ambient concentrations of ozone. Tropospheric ozone pollution occurs at ground level and is the major component of urban smog. Ozone is a secondary pollutant formed in the atmosphere by reactions of volatile organic compounds (VOCs) and NO_x in the presence of sunlight. The EPA established a national ambient air quality standard (NAAQS) for ozone in order to protect the public health and welfare. After two decades of efforts to reduce ozone concentrations, primarily through reductions in emissions of VOCs, tropospheric ozone remains a widespread and important problem. A recent study by the National Academy of Sciences and EPA concludes that NO_x control is necessary for effective reduction of ozone in many areas.

The Title I requirements are directed at all types of major stationary sources of NO_x. On a nationwide basis, stationary source NO_x emissions originate primarily from five types of sources: utility boilers, gas turbines, internal combustion engines, process heaters, and industrial boilers. Approximately 85 percent of stationary source NO_x emissions are accounted for by these sources, with utility boilers comprising almost 60 percent of the total stationary source emissions. Other source categories can be important in individual areas.

The primary purpose of the acid rain NO_x emission reduction program is to reduce the adverse effects of acidic deposition on natural resources, ecosystems, visibility, materials, and public health by substantially reducing annual emissions of NO_x. NO_x emissions are a principal acidic deposition precursor. Although sulfate deposition is considered to be the major contributor to long-term aquatic acidification, nitric acidic deposition plays a dominant role in the "acid pulses" associated with the fish kills observed during the springtime meltdown of the snowpack in sensitive watersheds.

The Title IV NO_x requirements are directed at coal-fired electric utilities. Electric utilities are a major contributor to NO_x emissions nationwide: in 1980, they accounted for 30 percent of total NO_x emissions and, by 1990, their contribution rose to 38 percent of total NO_x emissions. Approximately 80 percent of electric utility NO_x emissions come from coal-fired plants of the type addressed by section 407 of the CAA.

While the primary purposes of NO$_x$ control are described above, NO$_x$ emission reductions have other environmental benefits as well. For example, the atmospheric deposition of nitrogen oxides is a substantial source of nutrients that damage estuaries such as the Chesapeake Bay by causing algae blooms and severely depleting the oxygen content. NO$_x$ emissions can be an important factor in certain areas which have not attained the nitrogen dioxide or particulate matter NAAQS. Particulate nitrate also contributes to pollutant haze, reducing visibility. Acidic deposition and ozone contribute to the premature weathering and corrosion of building materials such as architectural paints and stones.

Overview of Title I of the CAA

The long history of the CAA extends back before 1970. A summary of significant events occurring during its development was previously published (1).

Title I of the CAA Amendments of 1990 contains many new and revised requirements for areas that have not attained the national ambient air quality standards (NAAQS) for ozone, carbon monoxide, particulate matter, sulfur dioxide, nitrogen dioxide, and lead. The EPA developed a guidance document, called the General Preamble to Title I (2), to assist States regarding the interpretation of the various provisions of Title I, as amended. A Supplement to the General Preamble was subsequently published (3) and provides guidance on implementation of Title I NO$_x$ provisions. This section of the paper focuses on the Title I requirements related to the ozone NAAQS.

Ozone Classifications. The CAA sets a new classification structure for ozone nonattainment areas based on the severity of the nonattainment problem. For each area classified under this section, the attainment date is as expeditious as practicable but no later than the date in the following table. The classification scheme is as follows:

Area classification	Design value, ppm	Primary standard attainment date
Marginal	0.121 up to (but not including) 0.138	November 15, 1993
Moderate	0.138 up to (but not including) 0.160	November 15, 1996
Serious	0.160 up to (but not including) 0.180	November 15, 1999
Severe	0.180 up to (but not including) 0.280	November 15, 2005
Extreme	0.280 and above	November 15, 2010

Additionally, a severe area with a 1986 to 1988 ozone design value of 0.190 up to, but not including, 0.280 parts per million (ppm) has 17 years (until November 15, 2007) to attain the NAAQS.

The designation/classification process for ozone was described in the Federal Register (4). For areas classified marginal to extreme, virtually all requirements are additive (e.g., a moderate area has to meet all marginal and moderate requirements, unless otherwise specified). The requirements reduce emissions of VOC and NO_x, which are precursors to ozone. This paper focuses on the NO_x requirements for major stationary sources.

State Implementation Plans (SIPs). The CAA requires EPA to establish NAAQS at levels which are requisite to protect public health. With the assistance of the States, EPA identifies areas that have attained the NAAQS, (attainment areas) and other areas which may need to reduce emissions in order to attain the NAAQS (nonattainment areas). The CAA gives States the primary responsibility for the attainment and maintenance of the NAAQS.

The CAA contains specific provisions for nonattainment areas and requires States to develop SIPs to implement these provisions to attain and maintain the NAAQS in those areas. The SIPs are air quality plans which include an analysis of the sources of pollution (emission inventory), a strategy to attain and maintain the NAAQS by reducing emissions from key sources (attainment or maintenance demonstration), and enforceable regulations to carry out the strategy.

Role of NO_x and VOC in Urban Ozone. Section 182(f) reflects a new directive in the 1990 amendments to the CAA that NO_x reductions are required in ozone nonattainment areas and transport regions, with certain exceptions. As a result, States are generally required to apply the same requirements to major stationary sources of NO_x as are applied to major stationary sources of VOC.

Over the last two decades, emphasis was placed on reducing VOC emissions in order to attain the ozone ambient air quality standard. The 1970 and 1977 amendments to the CAA did not explicitly require NO_x reductions from stationary sources for purposes of attainment of the ozone standard. The best available scientific evidence at the time suggested that VOC reductions were preferred in most instances. The VOC control approach was reinforced by the fact that NO_x reductions could in some cases increase ozone concentrations.

In the debate leading up to the 1990 CAA amendments, Congress included consideration of NO_x. A report by the Office of Technology Assessment (5) and cost analyses by EPA provided support for inclusion of cost-effective NO_x controls in the ozone program. The 1990 amendments changed the statutory framework to place NO_x reductions on a more equal footing with the VOC reductions.

In the process of adding these new NO_x requirements, Congress recognized that NO_x reductions would help achieve ozone reductions in some ozone nonattainment areas, but that "there are some instances in which NO_x reductions can be of little benefit in reducing ozone or can be counter-productive, due to the offsetting ability of NO_x to 'scavenge' (i.e., react with) ozone after it forms" (6). The Congress provided for additional review and study under section 185B of the CAA "to serve as the basis for the various findings contemplated in the NO_x provisions" (7). In discussing the new Title I NO_x provisions, the House Report also stated that the Committee (on Energy and Commerce) "does not intend NO_x reductions for reduction's sake, but rather as a measure scaled to the value of reductions for achieving attainment in the particular ozone nonattainment area" (8).

Under section 185B, the EPA, in conjunction with the National Academy of Sciences, conducted a study on the role of ozone precursors in tropospheric ozone formation which examined the role of NO_x and VOC emissions, the extent to which NO_x reductions may contribute or be counterproductive to achieving attainment in different nonattainment areas, the sensitivity of ozone to the control of NO_x, the availability and extent of controls for NO$_x$, the role of biogenic VOC emissions, and the basic information required for air quality models. The NAS portion of the study was published in 1992 (*9*). The section 185B study was completed and submitted to Congress July 30, 1993 (*10*).

The section 185B study concludes that, based on the latest scientific evidence, the ozone precursor control effort should focus on NO$_x$ controls in many areas and that the analysis of NO$_x$ benefits is best conducted through photochemical grid modeling. This apparent shift, explains the study, coincides with improved data bases and modeling techniques that provide the analytical means to evaluate the effectiveness of ozone precursor control strategies.

The New NO$_x$ Requirements. The CAA as amended in 1990 set new NO$_x$ requirements for major stationary sources under Title I. The new NO$_x$ requirements are reasonably available control technology (RACT) for existing major stationary sources of NO$_x$ and new source review (NSR) for new/modified major stationary sources. Title I also requires any additional control measures as needed to demonstrate attainment of the NAAQS. These NSR, RACT, and attainment demonstration requirements are summarized elsewhere in this paper and described in more detail in EPA's NO$_x$ Supplement to the General Preamble (*3*).

The NSR and RACT requirements apply to major stationary sources. The CAA's definition of major stationary source varies with the classification of the ozone nonattainment area. For marginal and moderate nonattainment areas a major stationary source for NO$_x$ is any stationary source that emits or has the potential to emit 100 tons per year or more of NO$_x$. In the case of serious, severe, or extreme ozone nonattainment areas within the transport region, lower threshold definitions of major stationary source apply to NO$_x$ sources (50, 25 and 10 tons per year, respectively).

These NSR and RACT requirements apply in certain ozone nonattainment areas (described elsewhere in this paper) and throughout an ozone transport region. The CAA (section 176A) allows the Administrator to establish an ozone transport region covering multiple States whenever interstate transport of pollutants contributes significantly to violations of the NAAQS. In the 1990 CAA amendments, Congress specifically created an ozone transport region comprising the States of Connecticut, Delaware, Maine, Maryland, Massachusetts, New Hampshire, New Jersey, New York,

Pennsylvania, Rhode Island, and Vermont, and the Consolidated Metropolitan Statistical Area that includes the District of Columbia.

The amended CAA requires States to adopt SIP revisions subject to EPA approval that incorporate the NSR and RACT requirements of the CAA. These new rules for ozone nonattainment areas were generally required to be submitted by November 15, 1992. States, in their RACT rules are generally expected to require final installation of the actual NO$_x$ controls by May 31, 1995 from those sources for which installation by that date is practicable. In some instances EPA would allow additional time for rule submittal and control installation (3).

NO$_x$ NSR Requirements. The CAA NO$_x$ provisions for NSR apply to all ozone nonattainment areas with a classification of marginal or higher. Previously these NO$_x$ NSR requirements applied only in areas designated nonattainment for the nitrogen dioxide NAAQS (i.e., only in the Los Angeles, CA area).

The NSR requirements include, but are not limited to, requirements that a new or modified major stationary source will apply controls representing lowest achievable emission rate (LAER) and that the source will obtain an emission offset prior to operation. LAER is generally defined as the most stringent emission limitation which is achieved in practice by such category of source. In many cases LAER would include consideration of catalytic reduction controls. Emission offsets refers to the CAA requirement that a new or modified major stationary source obtain emission reductions from the same source or other sources in the area so that there is no net increase in emissions in the area. Also, State rules must ensure that NO$_x$ offsets will be consistent with any applicable attainment demonstrations. The NSR requirements detailed in section III.G of the General Preamble (2) for major VOC sources also apply for major NO$_x$ sources.

NO$_x$ RACT.

The CAA NO$_x$ provisions for RACT apply in all ozone nonattainment areas classified moderate or higher and throughout an ozone transport region. The RACT provisions do not apply to marginal ozone nonattainment areas except in ozone transport regions.

General Definition of RACT. The EPA has defined RACT as the lowest emission limitation that a particular, existing major stationary source is capable of meeting by the application of control technology that is reasonably available considering technological and economic feasibility (*11*). Although EPA historically has recommended source-category-wide presumptive RACT limits, and plans to continue that practice, decisions on RACT may be made on a case-by-case basis, considering the technological and economic circumstances of the individual source.

RACT may require technology that has been applied to similar, but not necessarily identical, source categories. Presumptive RACT limits are based on capabilities which are general to an industry, but may not be attainable at every facility. An extensive research and development program should not be necessary before a RACT control technology can be applied to a source. This does not, however, preclude requiring a short-term evaluation program to permit the application of a given technology to a particular source.

RACT for Certain Electric Utility Boilers. As described in the NO$_x$ Supplement (*3*), the EPA has determined that, in the majority of cases, RACT will result in an overall level of control equivalent to the following maximum allowable emission rates (pounds of NO$_x$ per million Btu) for utility boilers:

(a) 0.45 for tangentially fired, coal burning;
(b) 0.50 for dry bottom wall fired (other than cell burner), coal burning;
(c) 0.20 for tangentially fired, gas/oil burning; and
(d) 0.30 for wall fired, gas/oil burning;

The EPA expects States, to the extent practicable, to demonstrate that the variety of emissions controls adopted are consistent with the most effective level of combustion modification reasonably available for its individual affected sources. Note, the coal-fired utility emission rates described above for Title I presumptive RACT are the same numeric emission rates as EPA set under Title IV.

The EPA encourages States to structure their RACT requirements to inherently incorporate an emissions averaging concept (i.e., installing more stringent controls on some units in exchange for lesser control on others). The State may allow individual owners/operators to have emission limits which result in greater or lesser emission reductions so long as the areawide average emission rates described above are met on a Btu-weighted basis.

RACT for Other Utility Boilers and Source Categories. For source categories and utility boilers other than the electric utility boilers specified above, EPA is not recommending a specific RACT level. In general, EPA expects that NO$_x$ RACT for these other sources will be set at levels that are comparable to the RACT guidance specified above for certain electric utility boilers. As described in a recent EPA guidance memorandum (12), comparability shall be determined on the basis of several factors including, for example, cost, cost-effectiveness, and emission reductions.

In general, the actual cost, emission reduction, and cost-effectiveness levels that an individual source will experience in meeting the NO$_x$ RACT requirements will vary from unit to unit and from area to area. These factors will differ from unit to unit because the sources themselves vary in age, condition, and size, among other considerations. The EPA's general RACT guidance urges States to judge the feasibility of imposing specific controls based on the economic and technical circumstances of the particular unit being regulated. In many cases, these factors are not the same in all States since the specific NO$_x$ RACT emission limitations and averaging times will differ from State to State. The EPA's presumptive NO$_x$ RACT levels for certain utility boilers are based on capabilities and problems which are general to the industry on a national basis. States may adopt statewide NO$_x$ RACT levels which are more stringent than the EPA levels based on statewide industry conditions. For these reasons, a single cost, emission reduction, or cost-effectiveness figure cannot fully describe the NO$_x$ RACT requirement.

Available data indicate that some coal burning wall-fired boilers can meet EPA's presumptive NO$_x$ RACT levels by application of low NO$_x$ burners at a cost effectiveness as low as $160 per ton. The data also indicate that certain tangentially-fired utility boilers may approach a cost-effectiveness level of $1300 per ton in order to meet the EPA presumptive NO$_x$ RACT levels.

In determining the NO$_x$ RACT comparable cost-effectiveness level, EPA believes that it is appropriate to focus on the range of cost effectiveness. The range is appropriate due to the variability of the actual cost effectiveness that is expected from unit to unit. Therefore, EPA has determined that NO$_x$ technologies with a cost-effectiveness range that overlaps the $160 to $1300 range should, at a minimum, be considered by States in the development of their NO$_x$ RACT requirements. Accordingly, application of certain technologies, such as selective catalytic reduction at utility boilers, does not appear necessary in order to meet the EPA presumptive NO$_x$ RACT levels.

In some cases, States will need to consider a broader cost-effectiveness range. For example, where States adopt NO_x RACT requirements that are more stringent than the EPA's presumptive RACT, the associated control technologies may result in higher cost-effectiveness figures and, thus, States should expect to apply a broader cost-effectiveness range. In addition, since the EPA's presumptive RACT levels are expected to be met by a majority of (but not all) sources, States should expect some sources to experience higher cost-effectiveness levels in order to meet the NO_x RACT requirements.

Attainment Demonstration. As described in Sections III.A.3 and III.A.4 of the General Preamble (2), States must provide a SIP for moderate and above classified ozone nonattainment areas that includes specific annual reductions in VOC and NO_x emissions as necessary to attain the NAAQS. This requirement supplements the RACT and NSR requirements described above. Thus, a State would need to require NO_x controls more restrictive than those provided by the NSR and RACT provisions where additional reductions in emissions of NO_x are necessary to attain the ozone standard by the attainment deadline.

This requirement can be met through applying EPA-approved modeling techniques described in the current version of EPA's Guideline on Air Quality Models. The Urban Airshed Model (UAM), a photochemical grid model, is recommended for modeling applications involving entire urban areas. In addition, for moderate areas contained solely in one State, the empirical model, city-specific Empirical Kinetic Modeling Approach (EKMA), may be an acceptable modeling technique. The State should consult with EPA prior to selection of a modeling technique. If EKMA is used, the attainment demonstration was due by November 1993; if UAM is used, the due date is November 1994.

Following application of the modeling that is required for moderate and above areas, a State must select and adopt a control strategy that provides for attainment as expeditiously as practicable. This decision must be addressed by a State whether or not an area was exempted from the November 1992 submittal of NO_x RACT and/or NSR rules and may result in revision of the previously adopted rules. In some instances the NO_x RACT and NSR requirements already adopted may need to be supplemented with additional or more advanced NO_x controls in order for the area to attain the NAAQS.

Advanced Control Technologies. In certain areas, States may require existing stationary sources to install NO_x controls based on "advanced control technologies"; i.e., control technologies that reduce emissions to a greater extent than either the RACT or Title IV requirements. For example, advanced controls would be required as part of a serious ozone nonattainment area's 1994 SIP if modeling found such controls to be necessary to provide for expeditious attainment of the ozone NAAQS. Advanced control techniques are also considered in the NSR program for major new or modified stationary sources. Advanced control technologies include, for example, selective catalytic reduction; this flue gas treatment technique uses an ammonia injection system and a catalytic reactor to selectively reduce NO$_x$ emissions and can be applied to several types and sizes of sources.

Exemption from the New NO$_x$ Requirements. Section 182(f)(1) states that the new NO$_x$ requirements shall not apply where any of the following tests is met:

(1)   in any area, the net air quality benefits are greater without NO$_x$ reductions from the sources concerned;

(2)   in a nontransport region, additional NO$_x$ reductions would not contribute to ozone attainment in the nonattainment area; or

(3)   in a transport region, additional NO$_x$ reductions would not produce net ozone benefits in the transport region.

In addition, section 182(f)(2) states that the application of the new NO$_x$ requirements may be limited to the extent necessary to avoid excess reductions of NO$_x$. Any person (or State) may petition the EPA to make the above findings.

The EPA's decision will be based on the documentation provided by the petitioner and application of the applicable EPA guidance (*13*). The EPA encourages the petitioners to consult with the appropriate State agencies and the EPA Regional Office during the development of the demonstration and plan revision to ensure that any exemption is likely to be approved and that any required rules can be adopted in a timely manner.

If EPA grants a petition, these Federal NO$_x$ requirements will no longer apply. However, States remain free to impose NO$_x$ restrictions on other bases. For example, States may choose in certain circumstances to reduce NO$_x$ emissions for purposes of ozone maintenance planning, visibility protection, PM-10 control strategy, acid deposition program or other environmental protection.

NO$_x$ Substitution. As provided in the CAA [section 182(c)(2)(C)], NO$_x$ control may be substituted for the VOC control required to meet the post-1996 VOC emissions reductions progress requirements or may be combined with VOC control in order to maximize the reduction in ozone air pollution for purposes of meeting those requirements. In order to substitute NO$_x$ reductions for VOC, the State must demonstrate to EPA, consistent with EPA guidance, that the NO$_x$ reductions would result in reductions in ambient ozone concentrations at least equivalent to that which would result from the amount of VOC emission reductions otherwise required.

In accordance with the applicable EPA guidance document (*14*), a State may demonstrate to the Administrator that the NO$_x$ substitution is justified. The EPA will make a formal determination on any State request when the Administrator approves a plan or plan revision. The EPA's decision will be based on the documentation provided by the State and application of the EPA guidance. The EPA encourages the States to consult with the appropriate EPA Regional Office during the development of the demonstration and plan revision to ensure that any such substitution is approvable and that any required rules can be adopted in a timely manner.

Technical Documents. The EPA has developed alternative control technique documents (ACT's) for seven source categories. The categories and examples of the control techniques contained in the documents are listed below:

1. Nitric and Adipic Acid Manufacturing Plants; extended absorption, selective catalytic reduction (SCR), and nonselective catalytic reduction (NSCR).

2. Stationary Combustion Gas Turbines; wet injection, dry low-NO$_x$ combustion, and SCR.

3. Process Heaters; low NO$_x$ burner (LNB), selective noncatalytic reduction (NSCR), and LNB + flue gas recirculation (FGR).

4. Stationary Internal Combustion Engines; air/fuel adjustment, ignition timing retard, prestratified charge, NSCR, and SCR.

5. Utility Boilers; low excess air + burners out of service (BOOS), LNB, LNB + advanced overfire air, SNCR, reburn, SCR, and FGR.

6. Industrial, Commercial & Institutional Boilers;

burner tuning/oxygen trim, water or steam injection, BOOS, LNB, overfire air, SNCR, reburn, SCR, and FGR.

7.   Cement Manufacturing; LNB, SNCR, SCR, and mid-kiln firing.

In addition, the following ACT documents are under development:

1.   Glass Manufacturing
2.   Iron and Steel

Through the ACT documents, EPA provides information on the full range of NO$_x$ control technologies for categories of stationary sources that emit or have the potential to emit 25 tons per year or more of NO$_x$. While the ACT documents will not contain presumptive RACT, they will contain extensive background information on control technologies, costs, etc., that can be used by States in making RACT determinations. Further, the ACTs can be used to identify advance control techniques that may be needed in a SIP attainment demonstration.

The EPA also maintains a clearinghouse of NO$_x$ control technology information from NSR permits and RACT rules. This information is available to the general public.

TITLE IV

Title IV of the 1990 Amendments provides for the reduction of NO$_x$ emissions from coal-fired utility boilers in two phases. In the first phase, two categories of burners are affected: dry bottom wall-fired and tangentially fired boilers (Group 1). Group 1 boilers under Phase I must meet certain performance standards by January 1, 1995. About one-quarter of all Group 1 boilers are covered in Phase I.

If more effective low NO$_x$ burner technology becomes available, EPA may promulgate more stringent standards by January 1, 1997, for Phase II dry bottom wall-fired and tangentially fired boilers. Such rulemaking would include NO$_x$ emission limitations for all other coal-fired utility boilers (Group 2) as well. However, Phase I units with Group 1 boilers will not be subject to any revised requirements. If new standards are not revised in 1997, Phase II units with Group 1 boilers will be subject, beginning January 1, 2000, to the emission limitations promulgated for the phase I program.

Emission Requirements. The EPA published a final NO$_x$ rule in the

Federal Register under the Title IV acid rain program (*15*). The final rule includes annual NO_x emission limitations of 0.50 lb/mmBtu for dry bottom wall-fired boilers and 0.45 lb/mmBtu for tangentially fired boilers.

The rule encourages early compliance with the Phase I, Group 1 standards by allowing Phase II units with Group 1 boilers that comply with the Phase I emission limitations by calendar year 1997, to be grandfathered from any revisions to the Group 1 standards until 2008. All other Phase II units will have to meet the revised standards in 2000.

The rule also establishes procedures allowing utilities with the same owner or operator, and the same designated representative, to average emissions among affected units to comply with the NO_x emission limitations. Further flexibility is provided by establishing procedures to allow affected units with Group 1 boilers to obtain an alternative emission limitation where it is demonstrated that they cannot meet applicable emission limitations through the use of low NO_x burner technology.

Applicability. The final rule applies to existing coal-fired utility units subject to SO_2 emission limitations or reduction requirements under Phase I or Phase II of the Acid Rain Program pursuant to sections 404, 405, and 409 of the Act, including substitution units designated and approved as Phase I units in substitution plans that are in effect on January 1, 1995. The rule also applies to new coal-fired units that are affected units allocated allowances under section 405 of the CAA.

The rule's definition of conventional, available low NO_x burner technology includes overfire air as an integral component. The purpose of section 407 of the CAA is the reduction of NO_x emissions to an average level set forth by Congress, and the most reasonable approach to achieving these reductions is through the flexible application of appropriate low NO_x burner technology. The approach taken by EPA in implementing the Congressional intent of these NO_x emission reductions has been to encourage a cost effective and judicious application of the level of low NO_x burner technology required to achieve the stated average annual emission levels.

Neither low NO_x burners nor overfire air are required to be installed on all units or on any particular unit. Consistent with the intent of section 407, the decision as to what level of control technology to install on any particular unit is left completely to the utility, based on the specific financial and operational needs of that utility. A reasonable and responsible utility will employ the full range of conventional and available low NO$_x$ burner technology components, including separated overfire air, in its response to the performance requirements set forth by Congress prior to applying for an exception to emit at a higher emission level. A unit that is unable to meet the applicable emission limitation using low NO$_x$ burner systems with air staging through the burner assembly only has several compliance options: (1) install a more effective NO$_x$ control technology (e.g., selective catalytic reduction) and meet the applicable limit; (2) install separated overfire air and apply for an AEL if the limit still cannot be met; or (3) to the extent it meets the requirements for averaging, participate in an averaging pool.

NO$_x$ RACT Relation to Title IV. All existing major stationary sources of NO$_x$, including coal-fired utility boilers, which are located in nonattainment areas must meet the NO$_x$ RACT and, if applicable, attainment demonstration requirements of Title I. In addition, coal-fired utility boilers must meet the NO$_x$ requirements of section 407 under the acid rain program in Title IV of the Act. Utilities should plan to meet the most stringent requirements applicable under the Act.

Summary

Title I of the CAA includes new NO$_x$ RACT and NSR requirements applicable to major stationary sources in certain ozone nonattainment areas and throughout an ozone transport region. Title IV of the CAA sets emission limits on coal-fired utility boilers in two phases. The requirements for NSR include installation of controls representing the lowest achievable emission rate and apply to major new stationary sources locating in ozone nonattainment areas; the NSR program is currently being implemented in ozone nonattainment areas. The NO$_x$ RACT/phase I of the Title IV NO$_x$ acid rain programs require certain existing major stationary sources to be retrofit with NO$_x$ controls; these controls are to be implemented in early 1995. Phase II of the acid rain program is due to be implemented by January 2000.

In addition to the above NO$_x$ requirements, Title I requires areas to develop and implement strategies sufficient to attain the ozone NAAQS in all areas by specific dates. Areas with more severe ozone problems are allowed more time to attain. These areas are in the process of completing dispersion modeling analyses which will lead to the adoption of strategies that provide for attainment of the ozone standard in the areas. The attainment strategies are due to be adopted in late 1994. Many of these attainment strategies will include new NO_x controls which achieve emission reductions much greater than the NO_x RACT/phase I acid rain limits. The new NO_x controls may require, for example, 60-80% reduction in NO_x emissions at utility boilers, as compared to 30-50% reduction which generally represents NO_x RACT for utility boilers. In other cases, States might determine, based on the dispersion modeling analyses, that NO_x reductions are not helpful and seek an exemption from any NO_x requirements. Additional dispersion modeling after 1994 might be conducted in many areas, especially those with attainment dates extending beyond 1999. These later analyses would reflect improvements in the dispersion models and advances in control technologies and could lead to revision of the attainment strategies.

The CAA NO_x requirements are new and experience with NO_x controls is limited. Not surprisingly, information on NO_x control options is now evolving rapidly. As States consider various attainment strategies and as facilities consider purchasing control equipment to meet the new State requirements, more information on control options will be needed. Control costs and reduction efficiencies in particular are key factors in decisions to adopt or purchase controls. In addition, other environmental and energy impacts of NO_x controls should be considered, such as the following: unit efficiency, e.g., boiler efficiency; emissions of pollutants other than NO_x, such as ammonia, carbon monoxide and hydrocarbons; ability to sell byproducts, such as flyash; and energy requirements which might have global warming implications.

As described in this paper, there is a clear need for the development of NO_x control systems to meet near- and long-term CAA requirements. The NO_x controls will help provide for attainment of the ozone NAAQS and will reduce the adverse effects of acidic deposition.

Literature Cited

1. U.S. Environmental Protection Agency, November 24, 1987, *Approval of Post-1987 Ozone and Carbon Monoxide Plan Revisions for Areas Not Attaining the National Ambient Air Quality Standards,* (52 *Federal Register* 45044).

2. U.S. Environmental Protection Agency, April 16, 1992, *General Preamble for the Implementation of Title I of the Clean Air Act Amendments of 1990*, (57 *Federal Register* 13498).

3. U.S. Environmental Protection Agency, November 25, 1992, *Nitrogen Oxides Supplement to the General Preamble; Clean Air Act Amendments of 1990 Implementation of Title I*, (57 *Federal Register* 55620).

4. U.S. Environmental Protection Agency, November 6, 1991, *Air Quality Designations and Classifications*, (56 *Federal Register* 56694).

5. *Catching Our Breath: Next Steps for Reducing Urban Ozone*, US Congress, Office of Technology Assessment, Washington D.C. 1989.

6. United States House of Representatives Report No. 490, 101st Congress, 2nd Sess., at 204.

7. United States House of Representatives Report No. 490, 101st Congress, 2nd Sess., at 257.

8. United States House of Representatives Report No. 490, 101st Congress, 2nd Sess., at 257-8.

9. *Rethinking the Ozone Problem in Urban and Regional Air Pollution*, National Research Council, National Academy Press, Washington D.C., 1992.

10. *The Role of Ozone Precursors in Tropospheric Ozone Formation and Control*, Office of Air Quality Planning and Standards, Environmental Protection Agency, Research Triangle Park, North Carolina, 1993; EPA-454/R-93-024.

11. U.S. Environmental Protection Agency, September 17, 1979, *General Preamble for Proposed Rulemaking on Approval of Plan Revisions for Nonattainment Areas--Supplement on Control Techniques Guidelines*, (44 *Federal Register* 53762).

12. Kent Berry, Office of Air Quality Planning and Standards, memorandum to the EPA Regional Office Air Directors on "Cost-Effective Nitrogen Oxides (NO_x) Reasonably Available Control Technology (RACT)," March 16, 1994.

13. John Seitz, Office of Air Quality Planning and Standards, memorandum and attached Guideline for Determining the Applicability of Nitrogen Oxides Requirements Under Section 182(f) to the EPA Regional Office Air Directors, December 16, 1993; and John Seitz, Office of Air Quality Planning and Standards, memorandum to the EPA Regional Office Air Directors, "Section 182(f) Nitrogen Oxides (NO_x) Exemptions--Revised Process and Criteria," May 27, 1994.

14. John Seitz, Office of Air Quality Planning and Standards, memorandum and attached NO_x Substitution Guidance to the EPA

Regional Office Air Directors, December 15, 1993.
15. U.S. Environmental Protection Agency, March 22, 1994, *Acid Rain Program; Nitrogen Oxides Emission Reduction Program*, (59 *Federal Register* 13539).

RECEIVED November 16, 1994

Chapter 3

Selective Catalytic Reduction of NO by NH₃ over Aerogels of Titania, Silica, and Vanadia

M. D. Amiridis[1], B. K. Na[2], and E. I. Ko[3]

[1]W. R. Grace & Co.–Connecticut, 7379 Route 32, Columbia, MD 21044
[2]Korea Institute of Science and Technology, Chemical Processes Lab,
P.O. Box 131, Cheongryang, Seoul, Korea
[3]Department of Chemical Engineering, Carnegie Mellon University,
Pittsburgh, PA 15213

Aerogels of titania, titania-silica, and titania-vanadia were prepared with titanium butoxide, silicon ethoxide, and vanadium triisopropoxide as precursors. The titania and titania-silica aerogels were then used as supports for vanadia, introduced by the incipient wetness impregnation of vanadium triisopropoxide and the subsequent calcination at 773 K. The structures of vanadia and titania in these samples were characterized by X-ray diffraction and Raman spectroscopy, and their catalytic properties by the selective catalytic reduction (SCR) of NO with NH₃. With H₂O and SO₂ in the feed stream, the SCR activity decreased with increasing vanadia loading. Samples of high activity all contained anatase TiO₂, but an active vanadia species only needed to be in close proximity and interacting with, and not necessarily deposited on the surface of crystalline titania. Co-gelling was thus an effective way to prepare an active sample in a single step, as demonstrated by the SCR data of the titania-vanadia aerogel. The addition of niobia, up to 10 weight %, did not appreciably change the surface area, structure, or SCR activity of the titania aerogel supported vanadia.

Titania-supported vanadia catalysts have been widely used in the selective catalytic reduction (SCR) of nitric oxide by ammonia *(1, 2)*. In an attempt to improve the catalytic performance, many researchers in recent years have used different preparation methods to examine the structure-activity relationship in this system. For example, Ozkan *et al. (3)* used different temperature-programmed methods to obtain vanadia particles exposing different crystal planes to study the effect of crystal morphology. Nickl *et al. (4)* deposited vanadia on titania by the vapor deposition of vanadyl alkoxide instead of the conventional impregnation technique. Other workers have focused on the synthesis of titania by alternative methods in attempts to increase the surface area or improve its porosity. Ciambelli *et al. (5)* used laser-activated pyrolysis to produce non-porous titania powders in the anatase phase with high specific surface area and uniform particle size. Solar *et al.* have stabilized titania by depositing it onto silica *(6)*. In fact, the new SCR catalyst developed by W. R. Grace & Co.-Conn., SYNOX™, is based on a titania/silica support *(7)*.

0097–6156/95/0587–0032$12.00/0
© 1995 American Chemical Society

Recently Handy and co-workers have used the sol-gel method to prepare gels containing various combinations of vanadia, titania and silica *(8, 9)*. Sol-gel synthesis is a versatile approach in preparing catalytic materials because it offers better control over their textural properties and sample homogeneity in multicomponent systems. Aerogels, materials that are obtained by drying wet gels with supercritical extraction, offer additional advantages such as high specific surface area and thermal stability *(10)*. We have explored some of these advantages by preparing aerogels of vanadia, titania and silica and evaluating their SCR activities. The specific issues that we have addressed are: (1) the use of a high-surface-area titania aerogel as a support, (2) the different support behavior between aerogels of titania and titania-silica, (3) the effect of having vanadia *in* rather than *on* titania, and (4) the effect of niobia addition. Our key findings are reported below.

Experimental

Sample Preparation. The preparation of the titania aerogel is described in detail elsewhere *(11)*. Briefly, we prepared a titania gel from a solution of methanol, titanium n-butoxide, nitric acid and doubly distilled water and subsequently removed the solvent by supercritical drying with carbon dioxide. To prepare a titania-silica aerogel containing 40 weight % (33 mol %) titania, we prehydrolyzed tetraethylorthosilicate (TEOS, same as silicon ethoxide) in a separate solution of methanol, nitric acid and water for 10 minutes before combining it with a solution containing the titanium precursor. A similar procedure, but without prehydrolysis, was used to prepare a titania-vanadia aerogel containing 10 weight % vanadia by using vanadium triisopropoxide as a precursor. All aerogels were calcined in flowing oxygen at 773 K for 2 h before the introduction of supported vanadia and niobia.

Four vanadia/titania samples, containing 4, 10, 15 and 20 weight % vanadia, were prepared by the incipient wetness impregnation of a titania aerogel with a solution of vanadium triisopropoxide in either methanol or 1-propanol. These samples were heated under vacuum at 383 K for 3 h, then at 573 K for 3 h and finally in flowing oxygen at 773 K for 2 h. The same sequence of heat treatment was used after the sample containing 4 weight % vanadia was impregnated with a methanol solution of niobium ethoxide. Three niobia-promoted samples, containing 2, 5 and 10 weight % niobia, were prepared. Table I summarizes all the samples used in this study. We use the notation A/B to denote that A is supported on B, and A-B to denote that A and B are mixed in the bulk.

Physical Characterization. The specific BET surface areas of all aerogels were determined from nitrogen adsorption data with a commercial Autosorb-1 instrument (Quantachrome Corp.). Powder X-ray diffraction patterns were obtained with a Rigaku D/Max diffractometer with Cu Kα radiation. Raman spectra were obtained with the 514.5-nm line of a Spectra Physics Model 2050-5W argon ion laser *(12)*. Samples were dried in an *in situ* cell at 383 K for 2 h in flowing oxygen before Raman measurements.

SCR Activity. In a typical run, 0.1 g of sample, after being pressed at 15,000 psi and sieved into 40/60 mesh particles, was tested in a stainless steel one pass flow reactor at atmospheric pressure, a temperature of 623 K and a flowrate of 200 l(STP)/h. The reactant concentrations were 400 ppm NO, 400 ppm NH₃, 4% O₂, 800 ppm SO₂, 8 % H₂O and the balance N₂. The NO concentration at both the inlet and the outlet of the reactor was analyzed by the use of a

Table I. Summary of Sample Characteristics

Notation	Description	BET Surface Area[1] (m²/g)	X-ray Diffraction Results[1]
4V/T	4 wt. % V$_2$O$_5$ on TiO$_2$ aerogel	156	anatase (s)[2]
10V/T	10 wt. % V$_2$O$_5$ on TiO$_2$ aerogel	86	anatase (s)
15V/T	15 wt. % V$_2$O$_5$ on TiO$_2$ aerogel	29	anatase (s), rutile (w), vanadia (w)[2]
20V/T	20 wt. % V$_2$O$_5$ on TiO$_2$ aerogel	30	anatase (s), rutile (w), vanadia (w)
4V/(T-S)40	4 wt. % V$_2$O$_5$ on TiO$_2$-SiO$_2$ aerogel containing 40 wt. % TiO$_2$	340	X-ray amorphous
10V/(T-S)40	10 wt. % V$_2$O$_5$ on TiO$_2$-SiO$_2$ aerogel containing 40 wt. % TiO$_2$	290	X-ray amorphous
(V-T)10	TiO$_2$-V$_2$O$_5$ aerogel containing 10 wt. % vanadia	133	anatase (s)
2Nb/4V/T	2 wt. % Nb$_2$O$_5$ on 4V/T	146	anatase (s)
5Nb/4V/T	5 wt. % Nb$_2$O$_5$ on 4V/T	141	anatase (s)
10Nb/4V/T	10 wt. % Nb$_2$O$_5$ on 4V/T	145	anatase (s)

[1] After calcination at 773 K for 2 h.
[2] (s) = strong; (w) = weak.

chemiluminescent analyzer (Thermo Electron).   Under these conditions diffusional limitations were important and therefore, standard correction methods *(13)* were employed in determining intrinsic rates. In particular, the model of Wakao and Smith *(14)* was used to calculate the effective diffusion coefficients. For similar catalysts, coefficients thus obtained were found to be within 10% of the experimentally measured values *(15)*. There was also some uncertainty associated with the catalyst particle size distribution (<5% contribution to the effectiveness factor uncertainty), and a 5-10% uncertainty in the conversion measurements. When all these factors are taken into account in a differential error analysis, they result in an overall uncertainty of approximately 15% for the intrinsic rates reported in this study.

Results and Discussion

Effect of Vanadia Loading. Figure 1 shows the Raman data for the four titania-supported vanadia samples; the region 700-1000 cm^{-1} is expanded in Figure 2 for clarity. For 4V/T and 10V/T, the major peaks are at 398, 518 and 634 cm^{-1}, corresponding to anatase titania, and at about 1030 cm^{-1}, corresponding to monomeric vanadia species that are tetrahedrally coordinated *(4, 16, 17)*. The 10V/T sample has an additional broad peak centered at ca. 940 cm^{-1} which is characteristic of a polymeric vanadia species *(4, 16, 17)*. At vanadia loadings of 15 and 20 weight %, vanadia crystallites can be identified by peaks at 284, 483, 700, and 993 cm^{-1} *(16, 18)*, together with rutile titania with peaks at 441 and 607 cm^{-1}. The peak positions are approximate because no attempt was made to deconvulute overlapping peaks. In terms of structural identification, these Raman results are in total agreement with the X-ray diffraction data summarized in Table I.

The fact that no vanadia crystallites are detected in 10V/T is significant because it demonstrates clearly the ability of a high-surface-area titania to stabilize dispersed vanadia species at high vanadia weight loadings. Bond and Tahir *(2)* suggested the monolayer capacity of titania-supported vanadia to be 0.10 weight % V$_2$O$_5$ per m^2 of support surface area. In other words, a support with a specific surface area of 100 m^2/g could stabilize 10 weight % of dispersed vanadia. Indeed, no vanadia crystallites were detected in our 10V/T sample which had a specific surface area of 86 m^2/g of sample, or 96 m^2/g of support. By comparison, when Went and co-workers *(16, 17)* deposited 9.8 weight % vanadia onto titania supports with surface areas of 50 m^2/g or less, their Raman data showed distinct vanadia crystallites.

The high surface area of the titania aerogel not only increases the monolayer capacity for vanadia, but also delays the formation of the polymeric vanadia species until higher vanadia weight loadings. For example, at a vanadia loading of 4 weight % the polymeric species is insignificant in our sample but accounts for close to half of the surface species in the samples examined by Went *et al.* *(16, 17, 19)*. In principle, our titania aerogel, which has a specific surface area of over 200 m^2/g before impregnation, has a monolayer capacity of 20 weight % vanadia. But this potential could not be realized because the introduction of 10 weight % or more vanadia and the subsequent heat treatment resulted in a significant loss of surface area.   Mercury porosimetry measurements indicate that larger pores (80-100 nm) are created in this case at the expense of the smaller ones (10-20 nm) which is consistent with a partial collapse of the gel structure.

Table II summarizes the SCR data for these four titania-supported vanadia samples.   The activities for 15V/T and 20V/T are similar on an overall mass and a vanadia content basis. These two samples contain mostly crystalline

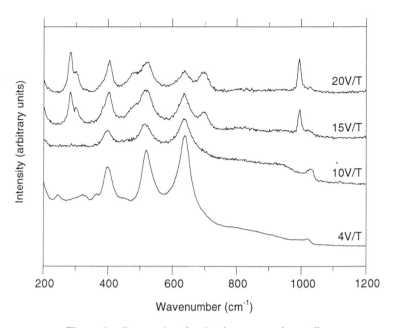

Figure 1.   Raman data for titania-supported vanadia.

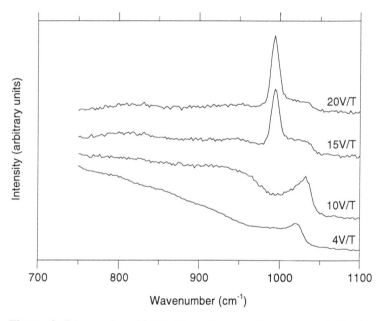

Figure 2. Raman data highlighting the vanadia region of titania-supported vanadia.

vanadia, which is less active than the dispersed species in 4V/T and 10V/T. In fact, since the Raman data suggest that the latter two samples contain only surface species, the most meaningful comparison of their activities is based on vanadia content. On that basis, 4V/T is about a factor of two more active than 10V/T, indicating a higher activity for the monomeric species. Went *et al. (19)* reported that the polymeric species is more active by almost an order of magnitude. Their experiments, however, were conducted at lower temperatures and in the absence of H$_2$O and SO$_2$. Thus, our results suggest that the variation of activity with vanadia loading may depend on whether H$_2$O and SO$_2$ are present in the feed stream.

Table II. Summary of SCR Data

		Intrinsic Rate	
Sample	Effectiveness Factor	(mol NO/g•s)	(mol NO/gV$_2$O$_5$ •s)
4V/T	0.54	1.0 x 10^{-5}	2.5 x 10^{-4}
10V/T	0.57	1.1 x 10^{-5}	1.1 x 10^{-4}
15V/T	0.74	4.7 x 10^{-6}	3.2 x 10^{-5}
20V/T	0.66	5.8 x 10^{-6}	2.9 x 10^{-5}
4V/(T-S)40	0.92	7.6 x 10^{-7}	1.9 x 10^{-5}
10V/(T-S)40	0.91	1.3 x 10^{-6}	1.3 x 10^{-5}
(V-T)10	0.27	2.0 x 10^{-5}	2.0 x 10^{-4}
2Nb/4V/T	0.29	1.1 x 10^{-5}	2.8 x 10^{-4}
5Nb/4V/T	0.31	1.3 x 10^{-5}	3.1 x 10^{-4}
10Nb/4V/T	0.35	9.7 x 10^{-6}	2.4 x 10^{-4}

Titania-Silica as a Support. Both X-ray diffraction and Raman data show no evidence of any crystalline phases of titania and vanadia in the 4V/(T-S)40 and 10V/(T-S)40 samples after calcination at 773 K. Apparently silica retards the sintering and crystallization of titania, as the support itself is X-ray amorphous prior to the introduction of vanadia. Furthermore, Raman data in Figure 3 show that, at comparable loadings of vanadia, the signal of surface vanadia species on titania-silica is a lot weaker than that on titania.

Researchers have discussed the need of crystalline titania in stabilizing active surface vanadia species for the SCR reaction *(8, 18)*. Handy *et al. (8)* showed that titania-silica mixed gels can be prepared to give either "titania-like" or "silica-like" behavior. Only on supports that contain crystalline titania domains can vanadia be dispersed to give high SCR activities. Even when silica-supported titania is used as a support, Jehng and Wachs *(16)* found an enhanced activity when bulk anatase particles are present. These authors suggested that interactions between surface vanadia species and bulk titania result in a more active site. In a related study, Deo *et al.* *(20)* showed that 1 weight % vanadia can be stabilized as a surface species on the anatase, rutile, brookite, and B phases of titania. It thus appears that as long as a crystalline titania phase is present, the specific crystal structure is of less importance in stabilizing an active surface vanadia species.

With the prehydrolysis of TEOS, we expect a homogeneous distribution of the two oxides in our titania-silica support. Coupled with the fact that silica is the major component, this sample should behave as the "silica-like" sample prepared by Handy *et al.* *(8)* with a two-stage hydrolysis procedure. In other words, there are no crystalline titania domains in the sample, consistent with the X-ray diffraction and Raman results. This titania-silica mixed oxide aerogel is therefore less effective in stabilizing surface vanadia species (see Figure 3) and less active in SCR (see Table II) than pure titania aerogel as observed.

Both Shikada *et al.* *(21)* and Bjorklund *et al.* *(22)* have prepared titania-silica, containing 50 mol % of each component, by co-precipitation and used it as a support for vanadia in SCR. But they do not report structural data for their supports, nor compare their activities directly with that of titania. We note that our 4V/(T-S)40 and 10V/(T-S)40 samples are also active for SCR. But without some form of crystalline titania, titania-silica is an inferior support compared to titania.

Effect of Vanadia in Titania versus Vanadia on Titania. Rate data in Table II show that the titania-vanadia aerogel, sample (V-T)10, is intrinsically more active than the titania-supported vanadia, sample 10V/T, on a vanadia content basis. This is despite the fact that in the case of (V-T)10 some of the vanadia probably remains in the bulk after heat treatment, and thus not available for catalysis. This result suggests that for an active SCR vanadia catalyst we may actually need vanadia species in close proximity with, and not necessarily supported on, crystalline titania.

Other researchers have prepared titania-vanadia mixed oxides by a sol-gel process *(9)*, a laser-induced process *(23)*, and a chemical mixing method *(24)*. Similar to our observations, in all these preparations titania crystallizes into the anatase phase upon calcination *(9, 23, 24)* and the mixed samples have comparable SCR activity to conventional impregnated titania catalysts *(9, 24)*. But it is unclear whether there is a structural difference in the active vanadia species in samples that are prepared differently. Our Raman data show that for the (V-T)10 sample, there is a peak at 1030 cm^{-1} corresponding to the monomeric vanadia species and no discernible peak at 940 cm^{-1} that would correspond to polymeric species. This could be due to the fact that vanadia is distributed homogeneously throughout the sample initially and thus does not aggregate upon heating. Recent evidence *(25)* suggests that in the co-gelling of two alkoxides, the more reactive component tends to form a "core" onto which the less reactive component attaches. Since the titanium precursor is more reactive than the vanadium precursor *(26)*, we would expect a titania "core" in this combination. This being the case, a co-gelled sample (e.g., our vanadia in titania) may be similar to a supported sample prepared conventionally (e.g., our vanadia on titania) in that one oxide is on or at the surface of another. The difference could very well be in the distribution and dispersion of the supported phase. Indeed, Handy *et al.* *(9)* noted a difference between their titania-vanadia mixed gel and vanadia salt-impregnated titania in that no evidence is found for a dispersed vanadia phase on the anatase crystallites in the former after heat treatment. Finally, Musci *et al.* *(23)* raised the interesting possibility that V atoms in a mixed oxide could induce strong variations in the electronic structures of titania and consequently alter catalytic properties. Even though

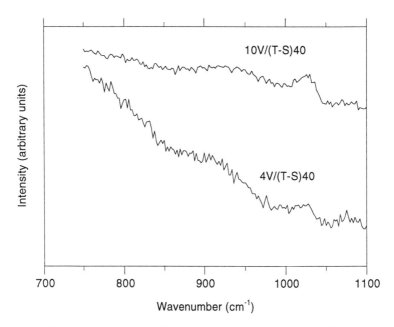

Figure 3.   Raman data for vanadia supported on titania-silica.

further work is necessary to resolve these issues, we conclude from our and others' work that different preparation methods have pointed to a subtle relationship between titania-vanadia interactions and SCR activity.

Effect of Niobia Addition. Table I shows that the addition of 2-10 weight % niobia to the 4V/T sample does not change its surface area or crystal structure. More importantly, the added niobia does not affect the structure of the dispersed vanadia species as shown by the Raman data in Figure 4. Thus, this series of samples allows us to examine the effect of niobia addition independent of other variables.

Rate data in Table II show that the addition of niobia did not significantly change the SCR activity of the 4V/T sample. If the monomeric vanadia species is indeed the active phase, then this finding is not surprising in view of the Raman results in Figure 4.   Apparently under our experimental conditions the niobia-vanadia interaction is weak and niobia itself does not have a direct role in the reaction.   This series of samples, however, may be of interest in other reaction systems.

Wang and Lee *(27)* recently reported the promoting effect of niobia on SCR activity.   Their finding is not inconsistent with ours because in their coprecipitated samples, niobia stabilizes the surface area upon heating, thereby increasing the acid and surface active sites. With our high-surface-area titania aerogel as a support, the addition of niobia does not bring about an advantage in surface area.

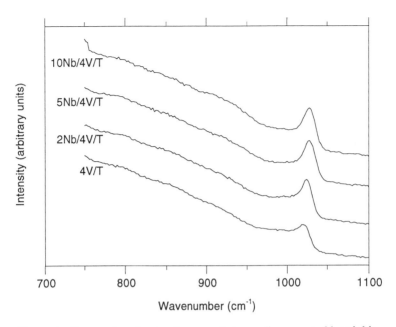

Figure 4.   Raman data for titania-supported vanadia promoted by niobia.

Conclusions

High-surface-area titania aerogels can stabilize dispersed vanadia species at higher vanadia loadings than conventional titania supports. Their full potential however, cannot be materialized because at vanadia loadings above 10 weight % a partial collapse of the gel structure and a subsequent loss of surface area take place upon heat treatment.

Vanadia species supported on titania aerogels demonstrate activity comparable to the activity of the same species on conventional titania supports for the selective catalytic reduction of NO by NH$_3$. Incorporation of vanadia in the gel structure results in an intrinsically more active material, which suggests that the active vanadia species need to be in close proximity and interacting with, but not necessarily deposited on the surface of, crystalline titania. On the contrary, the XRD amorphous titania-silica mixed aerogels, when impregnated with vanadia, were found to be less effective in stabilizing surface vanadia species and at least one order of magnitude less active for the SCR reaction than the pure titania aerogels. The addition of niobia, up to 10 weight %, to the titania aerogel supported vanadia does not significantly change the SCR activity under our experimental conditions.

Acknowledgment

We acknowledge W.R. Grace & Co. - Conn. for the permission to publish this work.

Literature Cited

1. Bosch, H. and Janssen, F., *Catal. Today*, **1988**, *2*, 369.
2. Bond, G. C. and Tahir, S. F., *Appl. Catal.*, **1991**, *71*, 1.
3. Ozkan, U. S., Cai, Y. and Kumthekar, M. W., *Appl. Catal. A*, **1993**, *96*, 365.
4. Nickl, J., Dutoit, D., Baiker, A., Scharf, U. and Wokaun, A., *Appl. Catal. A*, **1993**, *98*, 73.
5. Ciambelli, P., Bagnasco, G., Lisi, L., Turco, M., Chiarello, G., Musci, M., Notaro, M., Robba, D. and Ghetti, P., *Appl. Catal. B*, **1992**, *1*, 61.
6. Solar, J. P., Basu, P. and Shatlock, M. P., *Catal. Today*, **1992**, *14*, 211.
7. Hegedus, L. L., Beeckman, J. W., Pan, W. H. and Solar, J. P., U.S. Patent 4,929,586, May 29, 1990.
8. Handy, B. E., Baiker, A., Schraml-Marth, M. and Wokaun, A., *J. Catal.*, **1992**, *133*, 1.
9. Handy, B. E., Maciejewski, M. and Baiker, A., *J. Catal.*, **1992**, *134*, 75.
10. Ko, E. I., *Chemtech*, **1993**, *23*(4), 31.
11. Campbell, L. K., Na, B. K. and Ko, E. I., *Chem. Mater.*, **1992**, *4*, 1329.
12. Serghiou, G. C. and Hammack, W. S., *J. Chem. Phys.*, **1992**, *96*, 6911.
13. Beeckman, J. W. and Hegedus, L. L., *Ind. Eng. Chem. Res.*, **1991**, *30*, 969.
14. Wakao, N. and Smith, J.M., *Chem. Eng. Sci.*, **1962**, *17*, 825.
15. Beeckman, J.W., *Ind. Eng. Chem. Res.*, **1991**, *30*, 428.
16. Went, G. T., Oyama, S. T. and Bell, A. T., *J. Phys. Chem.*, **1990**, *94*, 4240.
17. Went, G. T., Leu, L. J. and Bell, A. T., *J. Catal.*, **1992**, *134*, 479.
18. Jehng, J. M. and Wachs, I. E., *Catal. Lett.*, **1992**, *13*, 9.
19. Went, G. T., Leu, L. J., Rosin, R. R. and Bell, A. T., *J. Catal.*, **1992**, *134*, 492.
20. Deo, G., Turek, A. M., Wachs, I. E., Machej, T., Haber, J., Das, N., Eckert, H. and Hirt, A. M., *Appl. Catal. A*, **1992**, *91*, 27.
21. Shikada, T., Fujimoto, K., Kunugi, T., Tominaga, H., Kaneko, S. and Kubo, Y., *Ind. Eng. Chem. Prod. Res. Dev.*, **1981**, *20*, 91.
22. Bjorklund, R. B., Andersson, L. A. H., Ingemar Odenbrand, C. U., Sjöqvist, L. and Lund, A., *J. Phys. Chem.*, **1992**, *96*, 10953.
23. Musci, M., Notaro, M., Curcio, F., Casale, C. and De Michele, G., *J. Mater. Res.*, **1992**, *7*(10), 2846.
24. Pearson, I. M., Ryu, H., Wong, W. C. and Nobe, K., *Ind. Eng. Chem. Prod. Res. Dev.*, **1983**, *22*, 381.
25. Miller, J.B., Rankin, S.E. and Ko, E.I., *J. Catal.*, accepted for publication.
26. Livage, L., Henry, M. and Sanchez, C., *Prog. Solid St. Chem.*, **1988**, *18*, 259.
27. Weng, R. Y. and Lee, J.F., *Appl. Catal. A*, **1993**, *105*, 41.

RECEIVED November 2, 1994

Chapter 4

Deactivation Behavior of Selective Catalytic Reduction DeNO$_x$ Catalysts

Basis for the Development of a New Generation of Catalysts

E. Hums and G. W. Spitznagel

Siemens AG, Power Generation Group (KWU), Freyesleben Strasse 1, 91058 Erlangen, Germany

Siemens Power Generation Group (KWU) has developed an SCR DeNOx plate-type catalyst without the use of Japanese catalyst technology. This innovative development is particularly suitable for demanding applications with flue gases from slag tap furnaces containing high heavy-metal concentrations (e. g. arsenic oxides). A special feature of this catalyst is that the high SO$_2$ content in flue gases from combustion of sulfur-rich fuel does not have any significant effects on DeNOx catalysis. The catalytically active phase used for this catalyst was developed in deactivation investigations on MoO$_3$-V$_2$O$_5$ compounds dispersed on TiO$_2$ (anatase). These investigations were conducted in the light of the high arsenic oxide concentrations in German slag tap furnaces with 100% ash recirculation.

The environmental impact of nitrogen oxides has focused attention on emissions regulations in many countries in recent years. The NOx emission limits imposed by German law cannot be achieved by simply applying primary measures such as staged combustion, over-fire air, etc.; this makes it necessary to apply secondary measures. Up to now, selective catalytic reduction (SCR) has dominated over other combustion control technologies.

A key reaction in the catalytic oxidation of ammonia in the presence of NO yielding N$_2$ and H$_2$O reflects aspects to the well-known Ostwald process for NO production when working unselectively (*1*). The advantages of this undesired side reaction were recognized and deliberately developed for oxidic catalysts in Japan, first employed in a coal fired power plant at Takehara in 1981 (*2*). These catalytic materials do not necessarily differ in composition from those which were used for NO production (*3*). Although this technique for removing NOx from waste gases using NH$_3$ as a reducing agent was referred to in an early Germany patent in 1963 (*4*); the first DeNOx catalytic reactor installed in Europe did not begin operation in Germany until late 1985 (*5*). In the USA, the technique is still in the pilot plant test phase (*6*).

A direct licensed transfer of the Japanese technology to German power plants with their specifications failed. This quickly became evident when Japanese catalysts tested in catalytic reactors downstream of steam generators with slag tap furnaces and ash recirculation exhibited drastic deactivation. This deactivation was caused by trace elements in the coals used, and affects catalyst lifetime (*7, 8*). This experience led to the initiation of extensive adaptation efforts, improving the catalyst material as well as engineering measures. Since no advanced technology was available in Japan at that time, Siemens began developing a plate-type catalytic converter in addition to licensed honeycomb-type fabrication.

The plate-type design exhibits advantages with regard to dust deposits, mechanical and thermal stability, low pressure drop, etc., which are described elsewhere (*9*). Therefore this design is primarily used in high-dust flue gas applications where the catalyst is exposed to high concentrations of heavy trace elements from the fired coals. These trace elements, which can act as catalyst poisons, are carried either in the flue gas and/or in the fine dust (*10*). Usually they cannot be prevented from depositing on the catalyst surface and they are expected to occur in much higher concentrations when ash is recirculated in the combustion chamber to achieve more complete combustion. This increases the As$_2$O$_3$ content by a factor of 20. Previous experience had shown that tungsten oxide-based catalysts deactivate at much faster rates in flue gases containing arsenic oxide than molybdenum oxide-based catalysts (*11*). These differences in behavior gave the impetus to clarify this phenomenon for the purpose of improving catalytic materials. Special attention is given to a TiO$_2$-Mo-V composite oxide catalyst, which shows quite different catalytic behavior and differences in other characteristics when compared with those catalysts obtained by monolayer dispersion of the oxides of vanadium and molybdenum onto TiO$_2$ (anatase).

Results and Discussion

Contamination and Deactivation of TiO$_2$-MoO$_3$-V$_2$O$_5$ Catalysts. When exposed to flue gas of coal-fired power plants with slag tap furnaces, TiO$_2$-MoO$_3$-V$_2$O$_5$ catalysts obtained by formation of a monomolecular dispersion of V$_2$O$_5$ and MoO$_3$ on TiO$_2$ (anatase) display an exponential correlation between exposure time and relative activity constant k/k$_0$. This correlation is shown in Fig. 1 for various power plants. In contrast to furnaces without ash recirculation, the influence of the deactivation process can already be detected after a very short period of operation (*12*).

Surface analysis of TiO$_2$-MoO$_3$-V$_2$O$_5$ Catalysts. This requires investigation of the degree of contamination of the catalyst surface as a function of time of exposure to flue gas, using X-ray photoelectron spectroscopy (XPS). Because of its detection depth of only a few atomic layers, XPS can yield information on surface contamination. The

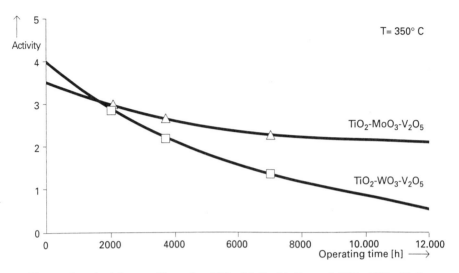

Figure 1. Activity profiles of a TiO$_2$-MoO$_3$-V$_2$O$_5$ and TiO$_2$-WO$_3$-V$_2$O$_5$ catalyst downstream of a slag tap furnace at T=350 °C.

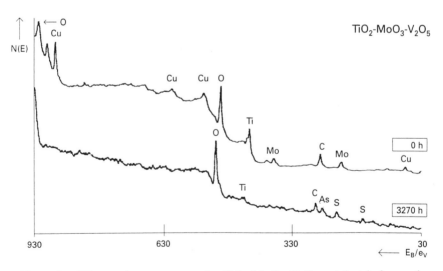

Figure 2. XP overview spectrum of a TiO$_2$-MoO$_3$-V$_2$O$_5$ catalyst before and after 3270 h exposure to flue gas of a slag tap furnace per 13.

detectability limit is better than 1%. In addition to a qualitative and quantitative analysis of the elements, their oxidation numbers should be focused on for further considerations. The calculated element ratio yields absolute values which allow a comparison between different samples. It was also possible to generate quantitative profiles enabling detection of dynamic changes in concentrations. All samples were exposed to the flue gas of a slag tap furnace with 100% ash recirculation. During the testing phase, no soot blowing took place and no NH_3 was added upstream of the sample. The intent was to study only the influence of the flue gas on surface composition as function of exposure time. Fig. 2 shows an overview spectrum of a TiO_2-MoO_3-V_2O_5 catalyst before and after 3270 h of flue gas exposure over the 500-1460 eV range (*13*). The samples were measured with Al K_α radiation. In addition to the main components of the catalyst, O, Ti, Mo and carbon introduced by catalyst preparation, and Cu and Fe from the sample holder and/or metallic catalyst support were detected. Apart from weak arsenic and sulfur signals, no further elements could be identified in this spectrum. However, increases in the As-Auger signals between 1210 and 1220 eV can be observed with increasing exposure time (Fig. 3). All samples contaminated with arsenic after power plant exposure show the clear doublet peaks in this region for the $L_3M_{45}L_{45}$ signal. The signals for arsenic were assigned to oxidation numbers +3 and +5 according to a reference sample which was obtained by a physical mixture of As_2O_3 and As_2O_5 and the catalyst before exposure to flue gas.

To quantify the elements deposited on the catalyst surface as a function of exposure time in the flue gas duct, their relative atomic fractions x_i were recorded with exposure time (Fig. 4). An increase in sulfur and arsenic contents with exposure time can be observed. Both elements were first detected after 48 h. Depending on exposure time, these elements cover increasingly large areas, 10% after 96 h, 20% after 3270 h. At the same time, all samples were observed to determine whether the detected X_{Ti} decreases with exposure time (Fig. 5). This reduction is particularly pronounced between 24 h and 48 h. After about 400 h of exposure time, the titanium fraction has decreased to about one third of its initial value. Since it does not change after this time, it is assumed that the catalyst surface contamination process is completed. From this time on, a constant balance should be achieved between abrasion and contamination. Surprisingly, this does not hold for sulfur and arsenic. It is assumed that the modified catalyst surface involved in this abrasion/contamination balance forms a further reactive component for these elements. There is other interesting evidence that the arsenic deposited originates, at least from that point on, exclusively in the gas phase, since the arsenic entrained by the flue dust would otherwise result in an increase in X_{Si}.

Poisoning Experiments on TiO_2-MoO_3-V_2O_5 Catalysts by As_2O_3.   If it is true that arsenic is mainly responsible for the deactivation phenomenon (*14-16*), an exponential correlation between arsenic oxide adsorption and the relative activity constant k/k_0 should be observed for the TiO_2-MoO_3-V_2O_5 catalyst when it is exposed to As_2O_3 doses in a laboratory test. Fig. 6 confirms such a dependency (*17*). Arsenic adsorption is not homogeneous through the entire catalyst, but steadily decreases from the outer surface to the center of the plate, as the concentration profile generated by SEM/EDX analysis shows (Fig. 7). This profile changes over exposure time in the flue gas. The arsenic concentration continues to increase in the inside of the

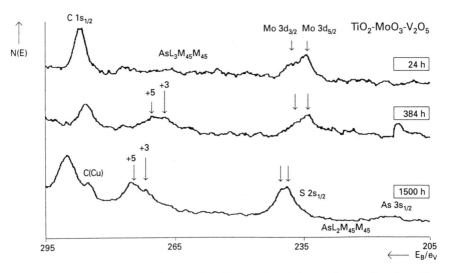

Figure 3. Arsenic Auger signals of a TiO$_2$-MoO$_3$-V$_2$O$_5$ catalyst as function of exposure time to flue gas of a slag tap furnace per 13.

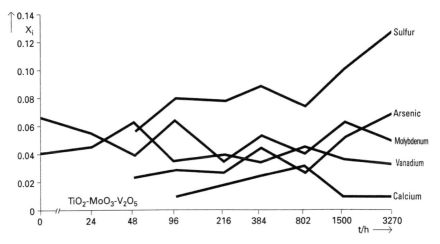

Figure 4. Atomic fractions X$_i$ of S, As, Mo, V, and Ca as a function of exposure to flue gas of a slag tap furnace (lines provided for clarity) of a TiO$_2$-MoO$_3$-V$_2$O$_5$ catalyst per 13.

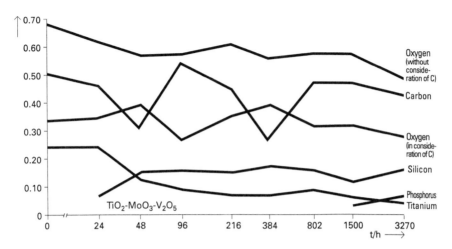

Figure 5. Atomic fractions X_i of O, C, Si, P and Ti as a function of exposure to flue gas of a slag tap furnace (lines provided for clarity) of a TiO$_2$-MoO$_3$-V$_2$O$_5$ catalyst per 13.

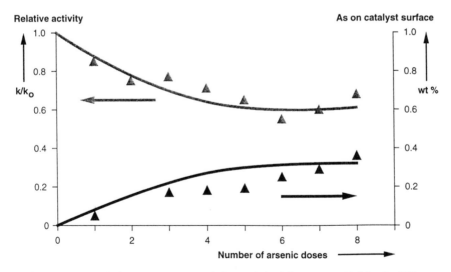

Figure 6. Correlation between the relative activity k/k$_o$ at T=350 °C of a TiO$_2$-MoO$_3$-V$_2$O$_5$ catalyst and the arsenic adsorbed in a laboratory experiment per 17.

catalyst, while it increases only slightly on the outer surface. In the power plant, the specific surface area of about 50 m^2/g decreases only slightly with the typical arsenic concentrations in the plant and loadings of up to 2 wt% of arsenic oxide. However, in the laboratory test, where the identical loading is reached in a shorter time with a much higher As concentration, the specific surface area decreases significantly (about 10-15%). The larger the initial specific surface area of the catalyst chosen, the larger the catalytic loss after flue gas exposure. Comparison of the catalyst integral pore volume before and after flue gas exposure described elsewhere (Ramstetter, A.; Schmelz, H.; Wilbert, E. unpublished data) indicates that the micropores are affected in particular. Based on known deactivation phenomena, this behavior demonstrates the loss of catalytic sites by poisoning and/or blocking. Since both effects can be superimposed on porous catalysts, the adjustment of inner surface of the catalyst was seen as a protective function for the arsenic-sensitive, catalytically active phase distributed in the micropores and macropores. Therefore, a compromise for TiO$_2$-MoO$_3$-V$_2$O$_5$ catalysts downstream of dry-bottom furnaces with a specific surface area of 75 m^2/g and an averages pore radius of 100 Å (*18*) should subsequently strive for a much smaller BET surface area and larger pore radii for implementation downstream of slag tap furnaces.

Nature of As$_2$O$_3$ Poisoned Species of TiO$_2$-MoO$_3$-V$_2$O$_5$ Catalysts. The observed macroscopic deactivation due to arsenic oxide demands an understanding on the molecular level. This understanding of mechanistic aspects can be used to estimate further potentials of catalyst development. On the one hand, this procedure naturally entails difficulties with regard to the concentration of the catalytic phase as a minor component of the catalyst. On the other hand, the interpretation of results obtained using analytical methods restricted either to the surface, subsurface or bulk properties often leads to false conclusions when the restrictions are not acknowledged.

Catalyst areas not accessible to XPS were studied by Hilbrig et al. (*19, 20*) using the XAS technique on TiO$_2$-MoO$_3$-V$_2$O$_5$ catalysts after exposure to flue gas. Hilbrig attributes the arsenic adsorption edge in the XANES range to pentavalent arsenic verified reference compounds. The chemical shift of the arsenic edge is localized between that of As^{5+} in Ag$_3$AsO$_4$ and As$_2$O$_5$. This is proven by XPS when the catalyst surface is mechanically removed and deeper layers of the bulk are exposed for analysis. Based on these results a reaction path obviously involves first deposition of As$_2$O$_3$ from the gas phase on the catalyst surface and subsequent oxidation to As^{5+} when penetrating the bulk. An important question must follow: which are the next neighbors to arsenic? Because this question is rather difficult to answer by analytical methods for such minor species as molybdenum and vanadium supported on the catalyst surface, no direct proof using XAS is available at this time. The formation of V-O-As is examined, reports in the older literature refer only to hydrated vanado (IV) and vanado (V) arsenates. The reduction of V$_2$O$_5$ to VO(AsO$_3$)$_2$ caused by As$_2$O$_3$ in the temperature range between 360-500 °C, described by Chernorukow et al. (*21*), should also occur in the presence of supporting TiO$_2$ (anatase) (*17*). As shown earlier (*22*), the formation of these compounds in the bulk can be suppressed if MoO$_3$ is added to V$_2$O$_5$. Powdered mixtures of MoO$_3$ and V$_2$O$_5$ were therefore melted in quartz ampules at 700°C. The mixing ratio was chosen based on the mixing ratio of industrial DeNOx catalysts. In this range, the phase diagram shows MoO$_3$ as the excess phase and a ß composite oxide phase, the stoichiometry of which is a subject of controversy (Fig. 8). A crystalline composite oxide phase involving As cannot be detected after treatment of molten mixtures of MoO_3 and V_2O_5 with As_2O_3. The reaction of pure MoO_3 with As_2O_3 at temperatures of about 350-480 °C forms the $As_4Mo_3O_{15}$ phase (*23*). This phase was identified on the basis of its X-ray diffraction pattern. Initial efforts to generate single crystals were unsuccessful. Only a fine crystalline powder was formed. Surprisingly, extended temperature treatment results in a phase transformation into a more stable $MoAs_2O_7$ (Fig.9). Its structure with mixed arsenic valencies (*24, 25*) would fully satisfy the existence of As^{5+} along with As^{3+} (*13*). The orthoarsenate in the $MoAs_2O_7$ is a tetragonal structure between two edge-linked molybdenum octahedra sharing three oxygen atoms (Fig. 10). As a first estimate, unsaturated MoO_5 would be suspected to be the unpoisoned Mo species in the catalyst as claimed by Hilbrig (*19*). Since the $MoAs_2O_7$ is primarily of two-dimensional structure, such a molecule could conceivably be a monolayer dispersion supported on TiO_2 (anatase). A comparison at the distances between As and O in this phase (*26, 27*) with those obtained with EXAFS by Hilbrig (*19*) on molybdenum-containing catalysts after flue gas exposure shows, with the exception of the last As-O distance, good agreement with the results gathered for As^{5+} (Table 1). More theoretical attempts to support $MoAs_2O_7$ have been started by computer simulation (Hums, E.; Spitznagel, G. W. unpublished data).

Since As^{5+} is found in all TiO_2-based catalysts following exposure to flue gas, it necessary to determine whether the $As_2O_3 \rightarrow As^{5+}$ reaction is strongly linked to TiO_2 (anatase). Hilbrig et al. (*19*) had measured exclusively As^{5+} in the XANES range when mixtures of As_2O_3 and TiO_2 (anatase) were ground with a mortar and pestle.

Performance of V-Mo Oxide Catalyst.   If the V-Mo-O phase instead of the single oxides of molybdenum and vanadium is dispersed onto TiO_2 (anatase), the catalysts obtained can clearly be distinguished in terms of their properties. This holds for the microstructure as well. An average pore radius about ten times larger than that of the TiO_2-MoO_3-V_2O_5 catalyst can be attained with the composite oxide catalyst (Fig. 11). Despite the larger pore radii, the arsenic oxide level after 1800 h exposure to flue gas is also unexpectedly higher than in the TiO_2-MoO_3-V_2O_5 catalyst. This is surprising, as, if it is assumed that the surface is covered with a monolayer by thermal spreading of arsenic oxide, the capacity of TiO_2 should be exceeded at 0.2 wt% As_2O_3 per m^2 and the catalytic activity should decrease to zero (Ramstetter, A.; Schmelz, H.; Wilbert, E. unpublished data). This would correspond to about 0.13 wt.% As_2O_3 per m^2 at 75 m^2/g specific surface area, and more than 50% of the TiO_2 surface would be covered by the monolayer formed. However the amount of arsenic oxide adsorbed by the TiO_2-Mo-V oxide catalyst in no way correlates with the anticipated deactivation rate.

The TiO_2-MoO_3-V_2O_5 catalyst studied by Hilbrig et al. (*19*), which is analogous to that formed by monomolecular dispersion of MoO_3 and V_2O_5 onto TiO_2 (anatase), was characterized with XRD and laser Raman spectroscopy in terms of loading. It was established that the intensities of the TiO_2 (anatase) signals at 400, 518 and 639 cm^{-1} are sensitive to loading with molybdate (Knözinger, H.; Mestl, G., unpublished data). The signal intensities decrease as loading increases and should decline to a value corresponding to monomolecular dispersion. With identical element composition, as is necessary for a monomolecular dispersion, this dependence is not detected for the TiO_2-Mo-V oxide catalyst. The signal position of TiO_2 (anatase)

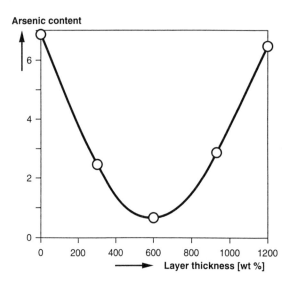

Figure 7. Graph of arsenic adsorbed over the layer thickness of a TiO$_2$-MoO$_3$-V$_2$O$_5$ catalyst measured with SEM/EDX per 17.

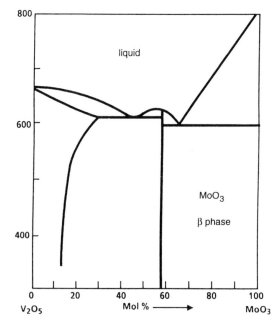

Figure 8. Simplified phase diagram of V$_2$O$_5$-MoO$_3$.

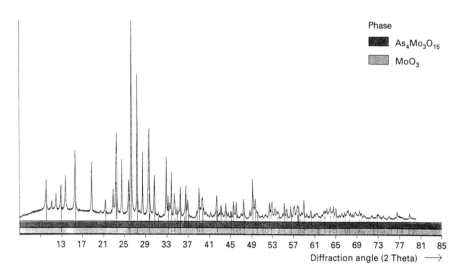

Figure 9. X-ray powder diffraction pattern of As$_4$Mo$_3$O$_{15}$ phase (1), MoO$_3$ phase JCPDS No. 35-0609 and residual pattern of phase transformation (1) → MoAs$_2$O$_7$ per 28.

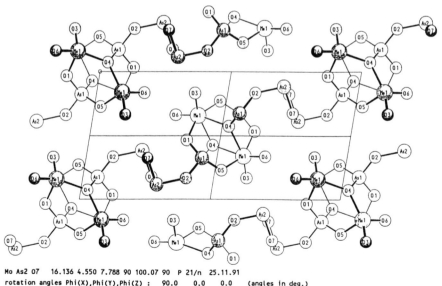

Figure 10. MoAs$_2$O$_7$ unit cell per 24.

Table 1. Calculated A-O distances for arsenic compounds.

Compound	As-0 Distances Å				Literature
TiO$_2$-MoO$_3$-V$_2$O$_5$ As^{5+}	1.67	1.67	1.67	1.94	Hilbrig, F. (1989)
LiMoO$_2$AsO$_4$ As^{5+} (3.95)* As^{5+} (4.41)*	1.681 1.702	1.688 1.689	1.689 1.665	1.720 1.717	Linnros, B. (1970)
MoAs$_2$O$_7$ As^{5+} (3.7)* As^{3+} (3.7)*	1.6708 1.7782	1.6720 1.7802	1.6835 1.8507	1.7011	Hums, E.; Burzlaff, H.; Rothammel, W. (1991)
As$_2$O$_3$ (4.0)*	1.78	1.78	1.78		Pertlik, F. (1978)

*) R-value for refinement

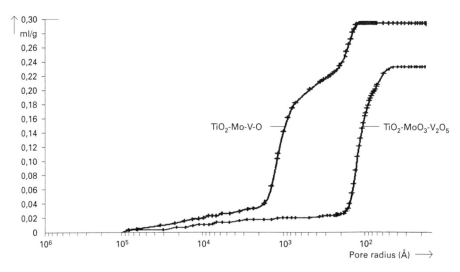

Figure 11. Comparison of pore distributions of a TiO$_2$-MoO$_3$-V$_2$O$_5$ catalyst and a TiO$_2$-Mo-V oxide catalyst per 11.

remains constant (Fig. 12), and after exposure to flue gas did not vary in position or intensity. This can be interpreted to mean that power plant exposure apparently does not influence dispersion quality.

Figure 13 shows k_{SO_x} as a function of k_{NO_x}, both measured at 350 °C for comparison of catalyst selectivity for the TiO$_2$-Mo-V oxide catalyst and the reference catalyst TiW and TiO$_2$-MoO$_3$-V$_2$O$_5$. If the SO$_2$-SO$_3$ oxidation rates of both catalyst types are compared, clear differences become evident. The TiO$_2$-Mo-V-oxide produces an SO$_2$-SO$_3$ oxidation rate which does not correspond to the vanadium content determined for the TiO$_2$-MoO$_3$-V$_2$O$_5$ catalyst. Whereas an increase in the vanadium content over 0.45 wt.% in the TiO$_2$-MoO$_3$-V$_2$O$_5$ catalyst results only in an increase in SO$_2$-SO$_3$ oxidation rates, this dependency cannot be demonstrated in the TiO$_2$-Mo-V oxide catalyst. Figure 14 shows catalytic activity as a function of plant operating time for various catalysts. The advantage of the TiO$_2$-Mo-V-oxide catalyst in terms of higher initial activity and lower deactivation rate over the other catalysts mentioned is clearly evident.

Conclusions

Investigations into the deactivation behavior of TiO$_2$-V$_2$O$_5$-MoO$_3$ catalysts have shown that the use of a V-Mo-oxide phase results in a DeNOx reaction rate which is many times higher than that obtained using the single oxides of vanadium and molybdenum. The formation of this phase is postulated to be the result of a phase transformation triggered by As$_2$O$_3$ or suitable substances on V$_2$O$_5$-MoO$_3$ mixtures. Results of investigations on surface and bulk specimens of V$_2$O$_5$-MoO$_3$ mixtures and the individual components of the TiO$_2$-MoO$_3$-V$_2$O$_5$ catalyst show that both systems can be clearly distinguished from each other.

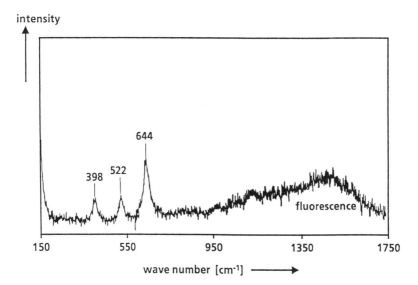

Figure 12. Laser Raman spectrum of TiO$_2$-Mo-V oxide catalyst after 1800 h exposure to flue gas of a slag tap furnace.

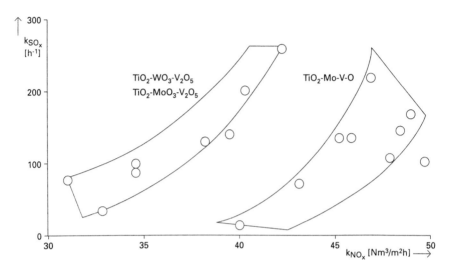

Figure 13. Comparison of the selectivity (k_{NOx}, k_{SOx}) of various catalysts at T=350°C.

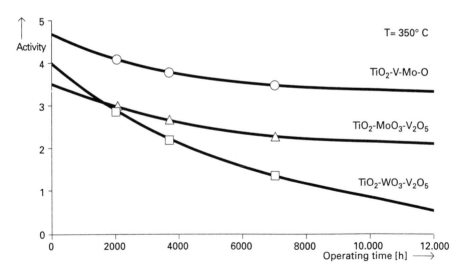

Figure 14. Activity profiles of a TiO$_2$-MoO$_3$-V$_2$O$_5$ and TiO$_2$-WO$_3$-V$_2$O$_5$ catalyst downstream of a slag tap furnace at T=350 °C.

Acknowledgements

We are grateful to the laboratory staff, in particular for the helpful support of Professor Wedler, Dr. Borgmann, Professor Burzlaff and Dr. Rothammel. Many thanks also to Professor Knözinger for recording a Raman spectrum of the catalyst material shown in this paper.

Literature Cited

1.   Mittasch, A. *Salpetersäure aus Ammoniak*; Verlag Chemie: Weinheim, Germany, 1953.
2.   Mori, T.; Shimizu, N. 1989 Joint Symposium of Stationary Combustion NOx Control, San Francisco, Session 6A: SCR Coal Applications.
3.   Ullmann, 1964; vol.15, pp 6-39; 1981; vol 20, pp 329-31.
4.   German Patent 1.253.685, 1968.
5.   Necker, P.; Becker, J. *VGB Kraftwerkstechnik* **1987**, *67*, issue 4, 368.
6.   Kuroda, H.; Morita, I.; Muxataka, T. 1989 Joint Symposium of Stationary Combustion NOx Control, San Francisco, Session 6A: SCR Coal Applications.
7.   Stäbler, K.; Schönbucher, B.; Bilger, H. *VGB Kraftwerkstechnik* **1988**, *68*, issue 7, 652.
8.   Schallert, B. *VGB Kraftwerkstechnik* **1988**, *88*, issue 4, 432.
9.   Hüttenhofer, K.; Spitznagel G. W., Pittsburgh Coal Conference 1993; Spitznagel, G. W.; Hüttenhofer, K.; Beer, J. K. ACS Symposium Series 552; 1994, pp 172-89; Spielmann, H.; Hüttenhofer, K. International Joint Power Generation Conference, Kansas City, 1993.
10. Braunstein, L.; Denzer, W.; Heß, K.; Schmidt, M. VGB-Konferenz "Chemie im Kraftwerk 1990" pp 80-4.
11. Hums, E. *Chemiker Zeitung* **1991**, *2*, 33.
12. Vogel, D.; Richter, F.; Sprehe, J.; Gajewski W.; Hofmann, H. *Chem.-Ing.-Techn.* **1988**, *60*, issue 9, 714.
13. Rademacher, I.; Borgmann, D.; Hopfengärtner, G.; Wedler, G.; Hums, E.; Spitznagel, G. W. *Surf. Interface Anal.* **1993**, *20*, 43.
14. Erath, R. *Techn. Mitt.* **1987**, *80* (9), 592.
15. Liebback, J. U.; Grünwald, K.-G.; Meyerhoff, M.; Otterstetter, H. *Energie Spektrum* **1987**, *6*, 50.
16. Gutberlet, H. *VGB Kraftwerkstechnik* **1988**, *68*, issue 3, 287.
17. Vogel, D. Dissertation, Erlangen 1989.
18. Schmelz, H. Keramische DeNOx-Katalysatoren, DKG-Fachausschuß 6, Höhr-Grenzhausen 1990.
19. Hilbrig, F. Dissertation, Munich 1989.
20. Hilbrig, F.; Göbel, H. E.; Knözinger, H.; Schmelz, H.; Lengeler, B. *J. Catal.* **1991**, *129*, 168.
21. Chernorukov, N. G.; Egorov, N. P.; Korshunov, I. A. *J. Russian Inorg. Chem.* **1978**, *23*, 1306.
22. Hums, E.; Göbel, H. E. *Ind. Eng. Chem. Res.* **1991**, *30*, 1814.
23. Hums, E; Göbel, H. E. *Powder Diffraction* **1990**, *5*, 170.
24. Hums, E.; Burzlaff H.; Rothammel, W. *J. Applied Catal.* **1991**, *73*, L19-L24.
25. Hums, E; Burzlaff, H.; Rothammel, W. *Acta Cryst.* **1993**, *C49*, 641.
26. Linnros, B. *Acta Chem. Scand.* **1970**, *24*, 3711.
27. Pertlik, F. *Czech. J. Phys.* **1978**, *B28*, 170.
28. Hums, E. *Ind. Eng. Chem. Res.* **1992**, *31*, 1030.

RECEIVED February 8, 1995

Chapter 5

Zeolite Selective Catalytic Reduction Catalysts for NO$_x$ Removal from Nuclear Waste Processing Plants

R. Gopalakrishnan and C. H. Bartholomew

Catalysis Laboratory, Department of Chemical Engineering,
Brigham Young University, Provo, UT 84602

NO$_x$ (typically 1-3%), present in the stack gas of nuclear waste process plants, is removed in the WINCO Process by two primary reactors to 300-1000 ppm by selective catalytic reduction (SCR) with NH$_3$ over a commercial zeolite catalyst at 300-500°C followed by reduction to low ppm levels in a third cleanup reactor. This study involved laboratory tests on advanced SCR zeolite catalysts, NC-301, ZNX, and Cu-ZSM-5, for the primary SCR reactors over a range of anticipated process conditions using gas mixtures containing 500-5000 ppm NO+NO$_2$, 500-5000 ppm NH$_3$, 1-2% CO, 14% O$_2$, and 20% steam in He. All three catalysts have acceptable levels of performance. i.e. selectively reduce >80% the NO$_x$ with NH$_3$ to N$_2$ over the temperature range of 400-500°C at a space velocity of 30,000 h^{-1}. The Cu-ZSM-5 catalyst is the most active and selective catalyst converting >95% NO$_x$ and NH$_3$ (at 500 - 5000 ppm of each) to N$_2$.

An interesting application of emissions control catalysts occurs in management and immobilization of spent radioactive fuels and wastes in the nuclear industry. During nuclear waste processing (NWP) at the Idaho Chemical Processing Plant (ICPP) NO$_x$ (NO and NO$_2$) and CO pollutants are typically discharged at levels of 1-3% to a waste gas stream. It is necessary to control both NO$_x$ and CO emissions in order to comply with current and anticipated regulatory requirements.

In earlier work, Thomas and coworkers (1-3) found that NO$_x$ in a simulated NWP offgas (containing 2% CO, 3% CO$_2$, 14% O$_2$, 20% H$_2$O and the remainder of N$_2$) could be removed from levels of 10,000 - 30,000 ppm to 300-1,000 ppm by selective catalytic reduction with ammonia over a commercial H-mordenite catalyst (Norton NC-300) at 300-500°C. Based on these results and pilot plant tests, a conceptual process for NO$_x$ and CO removal has been designed by Westinghouse Idaho Nuclear Company, Inc. (WINCO). This process consists of three fixed bed

0097–6156/95/0587–0056$12.00/0
© 1995 American Chemical Society

catalytic reactors (*4*); NO_x is removed in the first two beds to a level of 300-1000 ppm by selective catalytic reduction with NH_3 over a commercial H-Fe-mordenite catalyst (NC-300) at 300-500°C. A third reactor designed for clean-up of ammonia slip and unconverted CO will contain Pt and/or Cu based catalysts (*5*).

While the WINCO Process has been demonstrated on a small scale and promises to be effective on a large scale, there are economic incentives for development of more active catalysts having wider temperature windows for high conversion. Previous investigations by the BYU Catalysis Laboratory indicate that Cu-ZSM-5 is more active than H-mordenite for SCR of NO by NH_3 at low NO concentrations (*5, 6*). A new zeolite SCR catalyst, ZNX, available from the Engelhard Corporation is claimed to have high SCR activity and a wide operating temperature range. Norton has made modifications to their NC-300 mordenite catalyst to improve its thermal stability and conversion-temperature characteristics; these are incorporated in their NC-301 catalyst. In view of these important developments it would be desirable to examine these catalysts of potentially higher activity for the WINCO process.

The objective of this study was to investigate and compare the conversion vs. temperature behavior of promising high-performance zeolite catalysts, NC-301, ZNX, and Cu-ZSM-5, for NO_x reduction by NH_3 at conditions representative of the primary SCR reactors of the WINCO NWP SCR process.

Experimental

Materials. The commercial zeolite catalysts, NC-301 and ZNX, were obtained as pellets from Norton Co. and coated monolith from Engelhard Corp. respectively. Both catalysts were crushed and sieved to 0.2-0.6 mm size particles. The Cu-ZSM-5 powder used in this study was previously investigated for the SCR of NO by propane (*7*) and ammonia (*5*); it consisted of 1% Cu ion-exchanged into a ZSM-5 (Si/Al = 21) supplied by the PQ corporation.

NO and NO_2 (each >99% pure, 3000 ppm in He), NH_3 (>99.5% pure, 2000 ppm in He), and 10% CO in He were obtained from the Matheson gas company. Helium and oxygen tanks were purchased from Whitmore Oxygen Co., Salt Lake City, UT. Gases other than NO, NO_2, and NH_3 were dried with zeolite traps.

Procedure and Equipment. Reduction of NO_x (NO and NO_2) by NH_3 on NC-301, ZNX, and Cu-ZSM-5 zeolite catalysts was investigated at near atmospheric pressure over a temperature range of 150-550°C in a quartz flow microreactor. In a typical experiment, about 0.7-1.0 g of catalyst granules was used to obtain a space velocity of 30,000 h^{-1}. The reactor system (*5*) was equipped with an on-line Gas Chromatograph (Hewlett Packard 5890), a NO/NO_x analyzer (Rosemount Analytical, Model 951A), an NH_3 gas analyzer (Rosemount Analytical, Model 880) and an Allen CO/HC infrared analyzer. The analyzers and GC were calibrated every other day using appropriate calibration gases (NO, NH_3, CO, CO_2, N_2O, and N_2) obtained from the Matheson gas company. Concentrations of NO, NH_3, and CO were measured before and after reaction and conversions were plotted as a function of temperature. Concentrations of NH_3, NO, NO_2, and CO were measured by continuous analyzers to within ± 2-3%, while concentrations of N_2O and CO_2 could be analyzed by GC to within ±5%. However, measurement of the N_2 concentration by GC was accurate to only ±10%. The reactant gas mixture was bubbled through a water reservoir heated at a fixed temperature to maintain an appropriate concentration of water vapor. NH_3 was introduced just before the reactor cell. It was found that NH_3 was completely soluble in water, and hence none was detected in the gas phase. In this case, the dissolved NH_3 was estimated using an aqueous NH_3 electrode analyzer (Orion model 720A) by collecting water every 30 minutes. Significant amounts of nitrite/nitrate were observed in aqueous phase and their concentrations were measured using a nitrate electrode analyzer (Orion model 93-07).

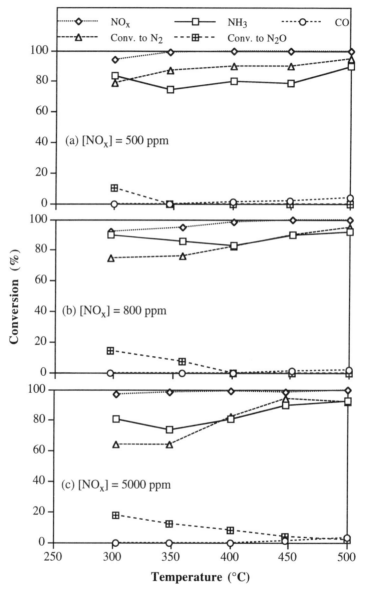

Figure 1. NO$_x$ reduction and CO oxidation conversion-temperature activities of NC-301 granules (0.2-0.6 mm size) as a function of inlet NO$_x$ concentration. [NO$_x$/NH$_3$] = 1, [NO$_2$/NO] = 2, [CO] = 1.5%, [O$_2$] = 14%, [H$_2$O] = 20%, and GHSV = 30,000 h^{-1}.

Results and Discussion

SCR Activities of NC-301, ZNX, and Cu-ZSM-5. SCR of NO_x by NH_3 on NC-301, ZNX, and Cu-ZSM-5 was carried out at different concentrations of NO_x and NH_3 (500, 800, or 5000 ppm, $NH_3/NO_x = 1$, and $NO/NO_2 = 0.5$) in the presence of 14% O_2, 1.5% CO, and 20% H_2O and at a space velocity of 30,000 h^{-1}. Conversions of NO_x, NH_3, CO, N_2, and N_2O were measured as a function of temperature and the results are presented in Figures 1-3.

At any given temperature, NO_x conversion is about 10-20% higher than NH_3 conversion, although the NO_x/NH_3 ratio in the feed is 1 (Figure 1). This is consistent with the observation of Thomas (*3*) that 99% of NO_2 conversion efficiency is obtained at sub-stoichiometric amounts of NH_3 at lower temperatures (300-350°C) on H-mordenite. According to previous studies of SCR of NO_x with ammonia on H-mordenite (*3, 8, 9*), possible stoichiometric reactions include:

$$6NO_2 + 8NH_3 \rightarrow 7N_2 + 12H_2O \tag{1}$$

$$8NO_2 + 6NH_3 \rightarrow 7N_2O + 9H_2O \tag{2}$$

$$5NO_2 + 2NH_3 \rightarrow 7NO + 3H_2O \tag{3}$$

$$6NO + 4NH_3 \rightarrow 5N_2 + 6H_2O \tag{4}$$

$$4NO + 4NH_3 + O_2 \rightarrow 4N_2 + 6H_2O \tag{5}$$

Thomas reported (*3*) that H-mordenite was more selective for NO_2 reduction in a mixture of NO and NO_2. If Equation 1 is a predominant reaction for NO_2, NO_x conversion is expected to be lower than NH_3 conversion when the NO_x/NH_3 ratio in the feed is 1. Moreover, if Equation 5 is the main route for NO reduction as suggested in previous work (*8*), NO and NH_3 conversions should be equal. Since, however, NO_x conversions are consistently higher than NH_3 conversions on NC-301 (Figure 1) and on H-mordenite (*3*), participation of Equations 2-4 would be consistent with the results. Thomas (*3*) attributed the observed higher NO_x conversion to Reaction 2, because of its large $\Delta H°$ of -53 kcal/mole. Indeed, in this study, significant N_2O formation was observed at temperatures below 400°C (Figure 1) indicating that Reaction 2 could be significant at lower temperatures; N_2O formation during SCR of NO_2 with NH_3 on H-mordenite was also observed below 400°C by Andersson et al. (*8*). Since N_2O is not observed at higher temperatures (>350°C), we conclude that N_2O reacts further to N_2 at high reaction temperatures. In fact, previous studies (*10-14*) indicate significant rates of N_2O decomposition on Fe-mordenite (*10*), H-mordenite (*11*), Fe-Y (*12*), cation-modified erionite (*13*), and cation-modified mordenite, ZSM-5, and beta zeolites (*14*) at temperatures of 300-500°C. Moreover, the NC-301 and ZNX catalysts of this study contain cation-exchanged Fe, while the Cu-ZSM-5 catalyst contains 1% ion-exchanged Cu.

Evidence for the participation of homogeneous or wall-catalyzed reduction of NO_2 to NO by Reaction 3 at temperatures above about 300-350°C was observed by Andersson et al. (*8, 9*) and by workers in this laboratory (*15*). Its contribution is typically on the order of 5-12% in the range of 350-500°C. Any NO produced by Reaction 3 would probably react readily via Reaction 5 to N_2 since Reaction 4 is negligible on H-mordenite (*16*). Since the overall stoichiometries of Reaction 3 and 5 predict higher NH_3 consumption relative to NO_2, this route would not account for the lower NH_3 conversion relative to that of NO_x.

NH_3 conversion of NC-301 increases from 84% to only a maximum of 90% as temperature increases from 300 to 500°C when the NH_3 feed concentration is 500 ppm (Figure 1a) suggesting that NH_3 oxidation may not be a significant reaction even at 500°C. An even smaller increase is observed when the NH_3 feed concentration is increased to 800 ppm or 5000 ppm (Figure 1b); i.e., conversions increase from 90 to 92% at 800 ppm and from 81 to 93% at 5000 ppm respectively when the temperature increases from 300 to 500°C. While NH_3 and NO_x conversions are generally greater than 80-90%, CO conversion reaches a maximum of only 2-4% even at 500°C,

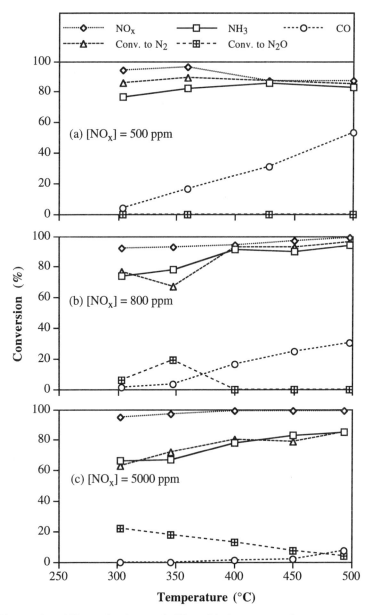

Figure 2. NOₓ reduction and CO oxidation conversion-temperature activities of ZNX/monolith granules (25 wt.% ZNX and 75 wt.% monolith) as a function of inlet NOₓ concentration. $[NO_x/NH_3] = 1$, $[NO_2/NO] = 2$, $[CO] = 1.5\%$, $[O_2] = 14\%$, $[H_2O] = 20\%$, and GHSV = 30,000 h^{-1}.

consistent with the earlier observation (5) that NC-301 is a poor CO oxidation catalyst.

SCR activities of ZNX and Cu-ZSM-5 were tested under the same conditions employed for NC-301, and the results are presented in Figures 2 and 3. Qualitatively, the conversion versus temperature behaviors of these catalysts are similar to NC-301; NO_x conversion is higher than NH_3 conversion, lower CO conversion is observed at higher initial NO_x concentration, and N_2O formation decreases with increasing temperature. However, quantitatively, they are different. Conversions of NH_3 and NO_x on ZNX are 77 and 94% respectively at 300°C (Figure 2a). While NH_3 conversion increases to 83%, NO_x conversion decreases to 87% as temperature is increased to 500°C; nevertheless, NO_x conversions are always greater than NH_3 conversions. These higher NO_x conversions relative to NH_3 conversions are similar to those for NC-301 suggesting that Reaction 2 contributes. At initial NO_x and NH_3 concentrations of 500 pm, CO conversion steadily increases from 4% at 304°C to about 53% at 500°C. As NO_x concentration increases from 500 to 800 ppm conversions of NH_3 and NO_x at 500°C increase to 94 and 99% respectively (Figure 2b), while CO conversion decreases to 30%. When the reactant concentration is further increased to 5000 ppm, NH_3 conversion decreases to 85%, NO_x conversion remains at 99% while CO conversion is decreased to 7% at 500°C (Figure 2c). This suggests that an increase in NH_3 and NO_x feed concentration has a small negative effect on NH_3 conversion, no effect on NO_x conversion but greatly decreases the overall CO conversion efficiency. Thus, ZNX appears to have SCR activity and selectivity performance comparable to that of NC-301, while having higher activity for CO conversion.

Conversions of NH_3, NO_x, and CO on Cu-ZSM-5 are 78, 97, and 12% respectively at 284°C with low inlet reactant concentrations, e.g., 500 ppm (Figure 3a). Conversions of NH_3 and CO increase to 99 and 91% while that of NO_x reaches 100% at 344°C; NO_x conversion decreases to 97% at 500°C. About 19% N_2O is produced initially at 284°C; this decreases to 0% at 416°C. The conversion trends for NO_x and NH_3 on Cu-ZSM-5 are similar to those of NC-301 and ZNX; i.e. NO_x conversions are higher than NH_3 conversions at temperatures below 400°C suggesting that a common SCR reaction mechanism (a combination of Equations 1-3) is operating on these three zeolite catalysts. Nevertheless, NH_3 conversion increases with temperature on Cu-ZSM-5 reaching ~100% at 450°C. The decrease in NO_x conversion at temperatures above 450°C is consistent with production of a significant amount of NO_x by NH_3 oxidation with molecular oxygen (5). When the NH_3 feed concentration is increased to 800 or 5000 ppm, NH_3 conversion increases from 86% at 310°C to 100% at 500°C, while the NO_x conversion pattern remains unaffected (Figure 3b). However, at any given temperature CO conversion is lower relative to the run with NO_x and NH_3 concentrations of 500 ppm each (Figure 3a); i.e. 70% at 495°C when the NO_x and NH_3 concentrations are 800 ppm (Figure 3b), while the CO conversion is only ~62% at 500°C when the NO_x and NH_3 concentrations are 5000 ppm (Figure 3c).

Effect of Monolith Addition on NC-301 and Cu-ZSM-5. Since only the ZNX catalyst was supplied in monolith form, tests were conducted with NC-301 and Cu-ZSM-5 physically mixed with crushed monolith. SCR of NO_x and CO oxidation activities were measured on a physical mixture of 25 wt.% NC-301 (0.2 - 0.6 mm size) with 75 wt.% of crushed monolith under the same conditions of Figure 1c (i.e. 5000 ppm NO_x, NO_x/NH_3 = 1, and NO/NO_2 = 0.5, 14% O_2, 1.5% CO, 20% H_2O, and 30,000 h^{-1} space velocity). The results (Figure 4) show that addition of 75% monolith by weight affects the catalytic performance of NC-301 only very slightly (i.e. slightly lower NO_x and NH_3 conversions at lower reaction temperatures); indeed the results are nearly the same within experimental error. The NO_x conversion activity of Cu-ZSM-5 (Figure 5) is not altered by the presence of monolith; however, NH_3 conversion is decreased nearly 10% at 300°C, while CO conversion is decreased from 62% at 492°C to 33% at 498°C by addition of monolith indicating that at higher

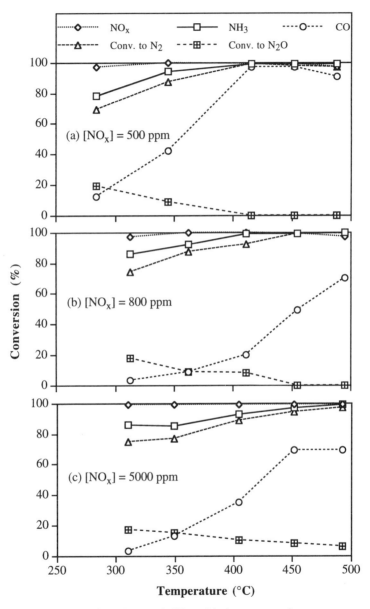

Figure 3.   NO$_x$ reduction and CO oxidation conversion-temperature activities of Cu-ZSM-5 powder as a function of inlet NO$_x$ concentration. [NO$_x$/NH$_3$] = 1, [NO$_2$/NO] = 2, [CO] = 1.5%, [O$_2$] = 14%, [H$_2$O] = 20%, and GHSV = 30,000 h^{-1}.

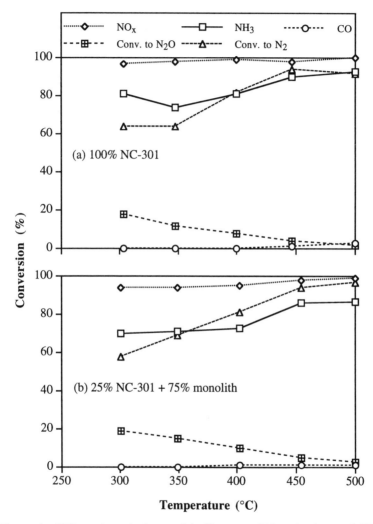

Figure 4. Effect of crushed monolith diluent on NO reduction and CO oxidation activities of NC-301 (0.2-0.6 mm size) catalyst. $[NO_x]$ = 5000 ppm, $[NO_x/NH_3]$ = 1, $[NO_2/NO]$ = 2, $[CO]$ = 1.5%, $[O_2]$ = 14%, $[H_2O]$ = 20%, and GHSV = 30,000 h^{-1}.

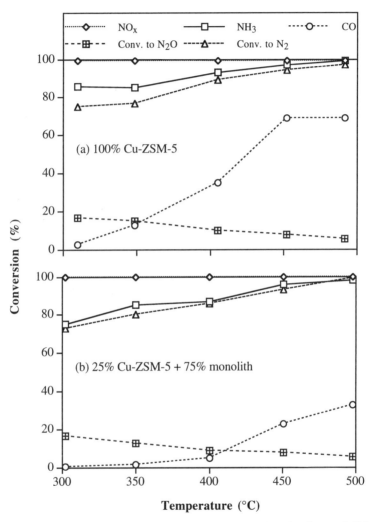

Figure 5.   Effect of crushed monolith diluent on NO reduction and CO oxidation activities of Cu-ZSM-5 catalyst. [NO$_x$] = 5000 ppm, [NO$_x$/NH$_3$] = 1, [NO$_2$/NO] = 2, [CO] = 1.5%, [O$_2$] = 14%, [H$_2$O] = 20%, and GHSV = 30,000 h^{-1}.

space velocity (or smaller catalyst charges) for a given concentrations of NO_x, NH_3, and CO, SCR of NO_x by NH_3 takes place preferentially over CO oxidation. It should be emphasized that the activity tests of NC-301 and Cu-ZSM-5 with crushed monolith not only provide a valid comparison with ZNX but also provide a reasonable measure of performance of these catalysts in wash coated-monolith form, since in either form the catalyst layer is roughly of the same thickness.

Comparison of Catalytic Performances of NC-301, ZNX, and Cu-ZSM-5. The results of catalytic performance of the three catalysts at 450°C are summarized in Table I. Figure 6 compares the conversions of NO_x, NH_3, and CO at 450°C on NC-301 (without monolith) as a function of inlet NO_x concentration (as always $[NH_3/NO_x] = 1$). NO_x conversion is apparently 100% at inlet NO_x concentrations of 500 and 800 ppm but decreases to 98% as the inlet NO_x concentration is increased to 5000 ppm. Therefore, about 100 ppm of NO_x slips through NC-301 at 450°C with an inlet concentration of 5000 ppm of NO_x. NH_3 conversion increases from 79% at 500 ppm NO_x to 90% at 800 ppm NO_x, and decreases to 86% at 5000 ppm NO_x. This indicates that NH_3 slip increases with increasing concentration of NH_3; for example, 55 ppm (11%), 80 ppm (10%), and 700 ppm (14%) of NH_3 slippage occurs at NH_3 inlet concentrations of 500, 800, and 5000 ppm respectively. CO conversion (1-2%) is, however, not affected by changes in NO_x concentration.

Similar analysis on ZNX (25 wt.% ZNX + 75 wt.% monolith) reveals that 75 ppm (15%), 16 ppm (3%), and 50 ppm (1%) of NO_x slippage and 65 ppm (13%), 80 ppm (10%), and 850 ppm (17%) of NH_3 slippage occur when using inlet NO_x and NH_3 concentrations of 500, 800, and 5000 ppm respectively (Table I and Figure 7). Significant CO conversion occurs on ZNX at low NO_x concentration but the conversion decreases from 37% to 2% as the NO_x concentration is increased from 500 to 5000 ppm.

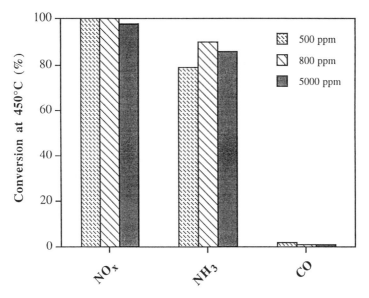

Figure 6. NO_x, NH_3, and CO conversions at 450°C on NC-301 as a function of NO_x concentration in the feed. $[NO_x/NH_3] = 1$, $[NO_2/NO] = 2$, $[CO] = 1.5\%$, $[O_2] = 14\%$, $[H_2O] = 20\%$, and GHSV = 30,000 h^{-1}.

Table I. Summary of catalytic performance of NC-301, ZNX, and Cu-ZSM-5 catalysts for removal of NO$_x$, NH$_3$, and CO at 450°C as a function of initial NO$_x$ concentration.

Properties	NC-301			ZNX[e]			Cu-ZSM-5		
	[NO$_x$] = 500 ppm[a,b]	[NO$_x$] = 800 ppm[a,b]	[NO$_x$] = 5000 ppm[c,d]	[NO$_x$] = 500 ppm[b]	[NO$_x$] = 800 ppm[b]	[NO$_x$] = 5000 ppm[d]	[NO$_x$] = 500 ppm[a,b]	[NO$_x$] = 800 ppm[a,b]	[NO$_x$] = 5000 ppm[c,d]
Conversion of									
NO$_x$	100	100	98	85	97	99	99	99	100
NH$_3$	79	90	86	87	90	83	99	99	96
CO	2	1	1	37	25	2	97	49	23
Conversion to									
N$_2$	90	90	94	87	93	79	98	99	93
N$_2$O	0	0	5	0	0	2	0	0	8

[a] Catalyst without monolith.
[b] Conversion to N$_2$ was calculated using N-material balance and assuming N$_2$ and N$_2$O are the only N-containing products.
[c] Catalyst with 75 wt.% monolith.
[d] Conversion to N$_2$ was estimated using GC analysis.
[e] The original catalyst contains 75 wt.% monolith.

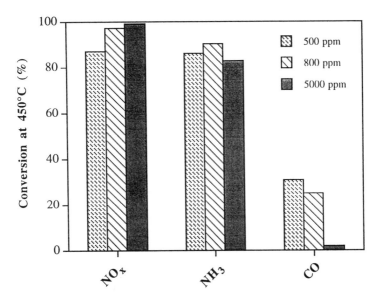

Figure 7. NO$_x$, NH$_3$, and CO conversions at 450°C on ZNX as a function of NO$_x$ concentration in the feed.   [NO$_x$/NH$_3$] = 1, [NO$_2$/NO] = 2, [CO] = 1.5%, [O$_2$] = 14%, [H$_2$O] = 20%, and GHSV = 30,000 h^{-1}.

Figure 8 compares the NO$_x$, NH$_3$, and CO conversions at 450°C on Cu-ZSM-5 as a function of NO$_x$ concentration. Apparently, negligible amounts of NO$_x$ slippage and only about 4% (200 ppm) of NH$_3$ slippage occur even at 5000 ppm of NO$_x$ (Table I) indicating that Cu-ZSM-5 is a more efficient SCR and NH$_3$ oxidation catalyst than NC-301 and ZNX.

The CO conversion versus temperature performances of NC-301, ZNX, and Cu-ZSM-5 are compared in Figure 9. In addition to the advantages of Cu-ZSM-5 in SCR and NH$_3$ oxidation reactions, its activity for CO oxidation is also higher than NC-301 and ZNX at all temperatures; for example, about 97% CO conversion is achieved on Cu-ZSM-5 in the presence of 500 ppm of each NO$_x$ and NH$_3$ compared to 2% and 37% CO conversions obtained on NC-301 and ZNX at 450°C. However, CO conversion decreases to 23% on Cu-ZSM-5 as the NO$_x$ concentration increases to 5000 ppm, which is nevertheless, much higher than the conversions obtained on NC-301 (1%) and ZNX (2%).

Conclusions

Laboratory tests were conducted on advanced SCR zeolite catalysts, NC-301, ZNX, and Cu-ZSM-5, candidate catalysts for the primary SCR reactors of a NWP plant over a range of anticipated process conditions using gas mixtures containing 500-5000 ppm NO+NO$_2$, 500-5000 ppm NH$_3$, 1-2% CO, 14% O$_2$, 20% steam in He. All three catalysts have acceptable levels of performance; selectively reducing over 80% NO$_x$ with NH$_3$ to N$_2$ over the temperature range of 400-500°C. The Cu-ZSM-5 catalyst is the most active and selective catalyst for conversion of NO$_x$ and NH$_3$ to N$_2$.

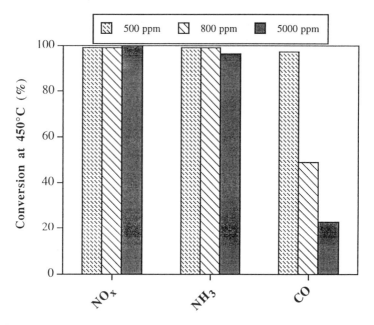

Figure 8. NO_x, NH_3, and CO conversions at 450°C on Cu-ZSM-5 as a function of NO_x concentration in the feed. $[NO_x/NH_3] = 1$, $[NO_2/NO] = 2$, $[CO] = 1.5\%$, $[O_2] = 14\%$, $[H_2O] = 20\%$, and GHSV = 30,000 h^{-1}.

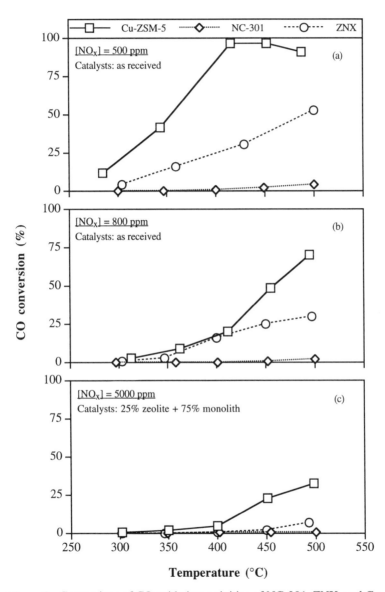

Figure 9. Comparison of CO oxidation activities of NC-301, ZNX, and Cu-ZSM-5 catalysts at different inlet NO_x concentrations. $[NO_x/NH_3] = 1$, $[NO_2/NO] = 2$, $[CO] = 1.5\%$, $[O_2] = 14\%$, $[H_2O] = 20\%$, and GHSV = 30,000 h^{-1}.

Acknowledgments

This work was supported by the Westinghouse Idaho Nuclear Company, Inc., the operating subcontractor of the Idaho Chemical Processing Plant at the Idaho National Engineering Laboratory, and also by the Advanced Combustion Engineering Research Center, Brigham Young University, Provo, Utah. Funds for this Center are received from the National Science Foundation, the State of Utah, 26 industrial participants, and the U.S. Department of Energy. Technical support by Mr. P. Stafford and Mr. S. Perry is also acknowledged.

References

1. Pence, D.T.; Thomas, T.R. *Proc. AEC Pollution Control Conf., Conf-721030,* October 25, **1972,** p.115.
2. Thomas, T.R.; Pence, D.T. *Proc. 67th Annual Meeting, Air Pollution Control Association,* June **1974,** pp. 75-258.
3. Thomas, T.R. "An Evaluation of NO_x Abatement by NH_3 over Hydrogen Mordenite for Nuclear Fuel Reprocessing Plant," *Report by Allied Chemical, Idaho National Engineering Laboratory to DOE, Contract EY-76-C07-1540,* January, **1978.**
4. Boardman, R. *Project Review Meeting; WINCO Research Needs,* Idaho Falls, May 11, **1992.**
5. Gopalakrishnan, R.; Davidson, J.; Stafford, P.; Hecker, W.C.; Bartholomew, C.H. in Environmental Catalysis, ACS Symposium Series **1994,** *552,* Chap. 7, p. 74.
6. Gopalakrishnan, R.; Hecker, W.C.; Bartholomew, C.H. *'Investigation of CO and NH_3 removal from ICPP Offgas,'* Final Technical Progress Report, submitted to WINCO, July **1993.**
7. Gopalakrishnan, R.; Stafford, P.; Davidson, J.; Hecker, W.C.; Bartholomew, C.H. *Applied Catal. B.* **1993,** *2,* 183.
8. Andersson, L.A.H.; Brandin J.G.M.; Odenbrand, C.U.I. *Catalysis Today* **1989,** *4,* 173.
9. Brandin, J.G.M.; Andersson, L.A.H.; Odenbrand, C.U.I. *Catalysis Today* **1989,** *4,* 187.
10. Leglise, J.; Petunchi, J.O.; Hall, W.K. *Journal of Catalysis* **1984,** *86,* 392.
11. Slinkin, A.A.; Lavrovskaya, T.K.; Mishin, I.V.; Rubinshtein, A.M. *Kinet. Katal.* **1978,** *19,* 992.
12. Sidamonidze, Sh.I.; Tsirsishvili, G.V.; Kheladze, T.A.; Sharabidze, L.M.; Soobshcheniya; *Akad. Nauk. Gruzinskoy SSR* **1987,** *3,* 605.
13. Aparicio, L.M.; Ulla, M.A.; Millman, W.S.; Dumesic, J.A. *Journal of Catalysis* **1988,** *110,* 330.
14. Li, Y.; Armor, J.N. *Applied Catalysis B: Environmental* **1992,** *1,* L21.
15. Bartholomew, C.H.; Gopalakrishnan, R. unpublished data **1994.**
16. Eng, J.; Bartholomew, C.H. unpublished data **1994.**

RECEIVED November 24, 1994

Chapter 6

Kinetic Studies of NO Reduction by CH_4 over Nonmetallic Catalysts

Xiankuan Zhang, Arden B. Walters, and M. Albert Vannice

Department of Chemical Engineering, Pennsylvania State University, University Park, PA 16802

NO reduction by CH_4 was conducted over MgO, 1% Li/MgO, 4% Li/MgO, 16% Li/MgO, and La_2O_3 between 773 and 973 K. The specific activities of N_2 formation (μmol N_2/s·m^2) over the Li/MgO samples were almost 5 times higher than that over pure MgO, and La_2O_3 was much more active than Li/MgO catalysts. E_{act} values for the Li/MgO catalysts were 29.0±1.1 and 35.0±1.0 kcal/mol in the absence and presence of O_2, respectively, while they were 24.4 and 26.0 kcal/mol for La_2O_3. Li/MgO had a negative reaction order of -0.5 in O_2 at 923K, whereas a positive 0.5 order in O_2 existed for La_2O_3. For the Li/MgO catalysts the reaction orders on CH_4 and NO were near 0.7 and 0.4, respectively, while they were about 0.3 and 1.0 for La_2O_3.

The reduction of NO_x (NO and NO_2) to N_2 is an important air emission control challenge for the process industries as well as for motor vehicles (1,2). Much recent study has been focused on the possibility that zeolite-based catalysts, in particular Cu/ZSM-5 (3-6) and Co/ZSM-5 (7-9), may be capable of selectively reducing NO_x with hydrocarbons. The degree of NO reduction very much depends on the reductants used (3). Hydrocarbons in the exhaust from a gas-cogeneration system, such as an electrical power plant, are comprised mainly of CH_4; therefore, it would be desirable to develop a catalyst which is active for the selective reduction of NO by CH_4 in oxygen-rich atmospheres. Recently Li and Armor have reported that NO reduction by CH_4 can effectively proceed on Co/ZSM-5 in the presence of excess oxygen (7,8). However, as observed with other zeolite catalysts, the activity of Co/ZSM-5 also showed a volcano-like dependence on temperature, with the maximum rate achieved around 700K. It is obvious that further investigations are needed not only to explore new catalyst systems which can effectively reduce NO with CH_4 at high temperature, but also to understand more of the fundamental chemistry associated with this reaction.

Lunsford and co-workers have shown that appropriate heterogeneous catalysts can generate methyl radicals at temperatures between 700 and 1000K (10-12). Our study is focused on testing a new concept — could CH_4 be activated on these oxidative coupling catalysts to efficiently reduce NO_x. Our investigation first examined NO

0097–6156/95/0587–0071$12.00/0

reduction by CH$_4$ over MgO and Li/MgO catalysts in the presence and absence of gas-phase O$_2$, then for comparison La$_2$O$_3$ was studied under identical conditions. Specific activities, activation energies, and reaction orders were obtained for all these catalysts.

Experimental

Li/MgO samples were prepared by following the recipe of Lunsford and coworkers (10,11) using MgO (Aldrich, 99.99%) and Li$_2$CO$_3$ (Aldrich, 99.997%), except for calcination at 973 K rather than 1023 K for 10 h under flowing dry air (50 cm^3/min). Three samples of Li/MgO were prepared containing 1.0, 4.0, and 16 wt % Li, while the MgO and La$_2$O$_3$ samples were prepared by calcining MgO powder (Aldrich, 99.99%) and La$_2$O$_3$ powder (Rhone-Poulenc, 99.99%) under the same conditions except a calcination temperature of 1023 K was used for La$_2$O$_3$.

Activity measurements and kinetic studies were made at atmospheric pressure in a quartz differential microreactor under steady-state reaction conditions. A typical gas mixture, unless otherwise specified, of 2% NO and 0.5% CH$_4$ was passed through the reactor, which contained 100-200 mg catalyst. Before any data were taken, the samples were pretreated in 9.8% O$_2$/He at 973K at a flow rate of 20 cm^3/min until no CO$_2$ was detected. All the gases used were UHP Grade from MG Ind except for the NO (99.0+%, N$_2$<1%, N$_2$O<0.5%) mixed in He (4.04%NO/He). During the Arrhenius runs, a period of 30 min on stream was allowed at each temperature before any gas sample was taken. An ascending-temperature sequence from 773 to 953K (for MgO and Li/MgO) or to 973 K (for La$_2$O$_3$) was usually followed by a descending-temperature sequence in order to detect any deactivation during these measurements. The conversions were generally kept below 20% to allow differential reactor operation. The kinetic studies in the presence of gas-phase oxygen were carried out by adding 9.8%O$_2$/He to the reactants while maintaining a constant GHSV for the total flow. The effect of O$_2$ on NO and CH$_4$ conversion over both Li/MgO and La$_2$O$_3$ catalysts was also examined from 773 to 953K (for MgO and Li/MgO) or to 973 K (for La$_2$O$_3$) by varying the O$_2$ concentration between 0 and 3.0%. Partial pressure dependency measurements were made at 923K at P$_{NO}$ = 15 Torr while the partial pressure of CH$_4$ was varied, while P$_{CH4}$ = 3.8 Torr was used while the NO partial pressure was varied. This pressure for NO was also used when the O$_2$ pressure was varied, but the CH$_4$ pressure was 25.6 Torr. Reaction orders were determined using a power rate law. The reactor effluent was analyzed with a Perkin-Elmer GC (Sigma-2B) equipped with a Carboxen^{1000} column (Supelco) and a P-E Nelson 1020S integrator. Specific activities for NO reduction to N$_2$ are expressed as the number of N$_2$ molecules produced per second per square meter of catalyst.

Chemisorption of NO on 4%Li/MgO and La$_2$O$_3$ was performed in a high vacuum adsorption system equipped with a precision pressure gauge. A base pressure of <10^{-6} Torr was routinely obtainable. Details regarding the system and measurements have been described previously (13). Prior to exposure to chemisorption, the catalyst samples were pretreated at 973 K in flowing 9.8%O$_2$/He for 30 min followed by evacuation at the same temperature for another 30 min before cooling to 300 K while evacuating. The irreversible NO uptakes at 300 K on these catalysts were employed to calculate adsorption site densities by assuming one NO molecule per site. This allowed turnover frequencies (TOF) (molecule reacted/site·s) to be calculated for NO conversion to N$_2$.

Results

Activity and Selectivity. NO reduction by CH$_4$ was studied over MgO, 1.0% Li/MgO, 4.0% Li/MgO and 16% Li/MgO between 773 and 953K using 2%NO,

0.5%CH$_4$ (balance He) at a GHSV of 3,000 h^{-1}. With all samples, NO conversion increased with reaction temperature up to 953K. Below 893K respective selectivities to N$_2$ and N$_2$O were typically ~60% and ~40%, but selectivity to N$_2$ rapidly increased at higher temperatures while that to N$_2$O simultaneously fell. As an example, Figure 1 shows both NO conversion and selectivities to N$_2$ and N$_2$O as a function of temperature over 4% Li/MgO. Over MgO selectivity to N$_2$O fell to almost zero above 873K while that to N$_2$ increased to ~100%, although overall rates of N$_2$ formation were lower. The effect of O$_2$ on NO reduction was investigated over the three Li/MgO catalysts at temperatures of 823, 873 and 923K, and similar results were observed over all three Li/MgO samples. As can be seen in Figure 2, NO conversion rapidly decreases as oxygen concentrations rise to 0.7% then it slows and finally approaches a stable level at higher O$_2$ concentrations. The corresponding CH$_4$ conversion shows an opposite trend as a function of O$_2$ concentration. The inhibiting effect of O$_2$ on the activity becomes less at high temperature. More complete results have been reported elsewhere (*14*).

The reduction of NO by CH$_4$ over La$_2$O$_3$ was studied from 773 to 973 K using 2%NO, 0.5%CH$_4$ in He at a GHSV of 8,700 h^{-1}. In these experiments N$_2$ and CO$_2$ were the only significant reaction products detected, although a tiny amount of N$_2$O was observed, because the H$_2$O peak in the gas chromatogram was not well resolved. The N and C mass balances were each within ±2.5%. As shown in Fig. 3, both NO conversion and rate of N$_2$ formation (μmol N$_2$/s·m^2) continuously increased with reaction temperature up to 973 K, the highest temperature employed in this work. Unlike zeolite-based catalysts, no activity maximum with increasing temperature was observed with La$_2$O$_3$. Except for a small amount of N$_2$O in the product stream, whose concentration was routinely near that of the N$_2$O impurity in the feed (ca. 100 ppm), N$_2$ was essentially the only N-containing compound observed as a product. Therefore, the net rate of N$_2$O formation was near zero and selectivity of NO reduction to N$_2$ is close to 100% on La$_2$O$_3$ over the entire temperature range tested. The effect of oxygen on activity and selectivity of NO reduction by CH$_4$ on La$_2$O$_3$ was subsequently tested. The experiments were performed from 773 to 973 K in the presence of 1.0% O$_2$ at a total GHSV of 17,400 h^{-1}, and the results are also presented in Figure 3. Again both NO conversion and the rate of N$_2$ formation increase with temperature. It clearly shows that the presence of oxygen greatly enhances NO reduction and the specific activities increase by a factor of about 5. Once again the amount of N$_2$O in the effluent was very small. Because of NO$_2$ formation downstream from the reactor due to the gas-phase reaction of NO and O$_2$, the selectivity of NO conversion to N$_2$ could not be quantitatively determined in these experiments. Additional details are provided elsewhere (*15*). In the presence of O$_2$, CH$_4$ reacts not only with NO, but combustion of CH$_4$ can also occur as follows:

$$CH_4 + 4NO \longrightarrow 2N_2 + CO_2 + 2H_2O \qquad (1)$$
$$CH_4 + 2O_2 \longrightarrow CO_2 + 2H_2O \qquad (2)$$

Figure 4 shows CH$_4$ conversion and selectivity over La$_2$O$_3$, i.e., the rate of reaction of CH$_4$ with NO versus the overall rate of CH$_4$ conversion (which includes combustion). As expected, CH$_4$ conversion increases with reaction temperature with an apparent activation energy near 38 kcal/mol at conversion below 30% and the rates of both N$_2$ formation and deep oxidation to CO$_2$ increase with temperature in the presence of excess O$_2$. The selectivity of CH$_4$ in the reaction with NO showed a decrease from ~40% to ~30% as temperature increased from 773 to 853 K, but the selectivities remained nearly constant above 850 K although CH$_4$ conversion increased rapidly.

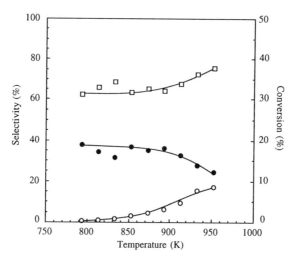

Figure 1. NO conversion (o) and selectivities to N$_2$ (□) and N$_2$O (●) over 4%Li/MgO (reproduced with permission from reference 14).

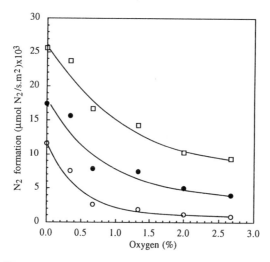

Figure 2. Effect of O$_2$ on N$_2$ formation over 4%Li/MgO at (o) 823 K, (●) 873 K and (□) 923 K (reproduced with permission from reference 14).

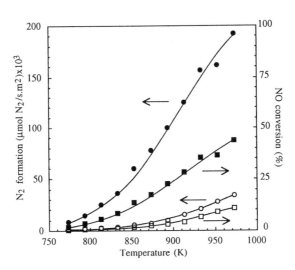

Figure 3. Rates of N₂ formation and NO conversion over La₂O₃ as a function of temperature: (o) and (□) 2%NO, 0.5%CH₄, balance He and GHSV = 8,700 h⁻¹, and (●) and (■) 2%NO, 0.5%CH₄, 1%O₂, balance He and GHSV = 17,400 h⁻¹ (reproduced with permission from reference 15).

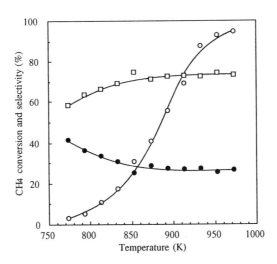

Figure 4. CH₄ conversion (o) and selectivities for CH₄ reacting with NO (●) and CH₄ reacting with O₂ (□) over La₂O₃. Reaction conditions: 2%NO, 0.5%CH₄, 1%O₂, balance He and GHSV = 17,400 h⁻¹ (reproduced with permission from reference 15).

Table I. Comparison of rates of N$_2$ formation over different catalysts

Catalyst	Reaction conditions				NO conv.	Rate of N$_2$ formation		TOF
	NO (%)	CH$_4$ (%)	O$_2$ (%)	Temp (K)	(%)	μmol/s·gcat x10^2	μmol/s·m^2 x10^3	(s^{-1}) x10^3
MgO	2.02	0.505	0	773	0.462	0.112	0.113	---
	2.02	0.505	1.0	773	0.375	0.091	0.091	---
1%Li/MgO	2.02	0.505	0	773	0.582	0.120	0.453	---
	2.02	0.505	1.0	773	0.199	0.041	0.151	---
4%Li/MgO	2.02	0.505	0	773	0.775	0.180	0.500	0.67
	2.02	0.505	1.0	773	0.364	0.085	0.236	0.32
16%Li/MgO	2.02	0.505	0	773	0.331	0.093	0.491	---
	2.02	0.505	1.0	773	0.205	0.058	0.304	---
La$_2$O$_3$	2.02	0.505	0	773	0.956	0.513	1.51	0.39
	2.02	0.505	1.0	773	1.93	2.89	8.50	2.21
	0.0820	0.0820	0	773	1.17	0.036	0.11	0.03
	0.0820	0.0820	0.5	773	1.72	0.21	0.61	0.16
Co/ZSM-5[a]	2.02	0.505	0	773	11.1	33.2	1.72	0.59
	2.02	0.505	0.5	773	42.6	128	6.63	2.29
Co/ZSM-5[b]	0.0820	0.0820	0	673	17	1.29	---	---
	0.0820	0.0820	0.5	673	60	4.56	---	---

[a]This sample was supplied by Air Products, and its BET surface area was measured by the authors.
[b]Ref. (7).

Table I lists NO conversions (to any and all products) and rates of N$_2$ formation obtained for these catalysts as well as TOFs of NO reduction to N$_2$ on some of the samples. The rates of N$_2$ formation in terms of μmole N$_2$/s·m^2 on the Li/MgO catalysts are almost 5 times higher than that on pure MgO. Clearly the presence of lithium promotes NO reduction; however, rates of N$_2$ formation are not a strong function of lithium loading. In fact, very similar results were observed with all three Li/MgO samples. Under identical conditions, La$_2$O$_3$ is much more active than MgO and 4%Li/MgO, especially in the presence of O$_2$; for example, the rates of N$_2$ formation (μmol/s·m^2) over La$_2$O$_3$, MgO and 4%Li/MgO under 2.02% NO and 0.505% CH$_4$ at 773 K in the absence of oxygen were 1.51, 0.11 and 0.50 μmol/s.m^2, respectively. Co/ZSM-5 is much more active than La$_2$O$_3$ on a weight basis, but when the rates are normalized to surface area, the activities of Co/ZSM-5 and La$_2$O$_3$ are close, as the measured BET surface area of the Co/ZSM-5 used in this work was 193 m^2/g. When compared in terms of TOF based on irreversible NO uptakes on 4%Li/MgO, La$_2$O$_3$, and Co/ZSM-5 at 300 K of 2.6, 13.1, and 560 μmol/g, respectively, the specific activities of 4%Li/MgO, La$_2$O$_3$, and Co/ZMS-5 are very close in the absence of O$_2$, while La$_2$O$_3$ remains comparable to Co/ZMS-5 with O$_2$ present. Both are more active than 4%Li/MgO in the presence of O$_2$. Table 1 also shows that the presence of oxygen enhances NO reduction over La$_2$O$_3$, whereas it has an inhibiting effect on NO reduction over MgO and Li/MgO catalysts. It should be pointed out that the reaction temperature of 673 K for Co/ZMS-5 is that where the maximum activity was observed (7), while that of 773 K for MgO, Li/MgO, and La$_2$O$_3$ was the lowest temperature employed in this work and corresponded to the lowest activities.

The rates of direct NO decomposition over MgO and Li/MgO catalysts were negligible, whereas direct NO decomposition occurred over La$_2$O$_3$ and it increased with temperature with an apparent activation energy of 23.0 kcal/mol. However, the rates of N$_2$ formation from direct NO decomposition can account for no more than 20% of the total N$_2$ formed during NO reduction by CH$_4$.

Table II. Kinetic behavior for NO reduction by CH$_4$

Catalyst	O$_2$ (%)	E$_a$ (kcal/mole)	Reaction Orders in		
			CH$_4$	NO	O$_2$[a]
MgO	0	27.6	0.60	0.44	---
	1.0	33.9	---	---	---
1%Li/MgO	0	29.4	0.73	0.41	---
	1.0	36.6	---	---	-0.73
4%Li/MgO	0	30.5	0.70	0.35	---
	0	28.7[b]	0.73[c]	0.48	---
	1.0	35.0	---	---	-0.43
16%Li/MgO	0	28.4	0.91	0.45	---
	1.0	34.5	---	---	-0.47
La$_2$O$_3$	0	24.4	0.26	0.98	---
	0	24.2[b]	---	---	---
	1.0	26.0	---	---	0.50

[a]O$_2$ concentration was below 2% over MgO and Li/MgO samples; and at higher O$_2$ concentration, the reaction order was zero.
[b]Reaction conditions: 0.2% NO, 0.5% CH$_4$, He balance and 1 atm total pressure
[c]NO was kept at 1.6 torr (0.2%)

Kinetic Studies. All the Arrhenius plots showed reproducible rates of N$_2$ formation in ascending- and descending-temperature sequences, and the relevant kinetic parameters are summarized in Table II. The apparent activation energies are 29.0 ± 1.1 kcal/mole over the Li/MgO catalysts, and no correlation between activation energies and lithium loadings is observed. Pure MgO has a very similar apparent activation energy although its activity is lower than those of the Li/MgO samples. The empirical reaction order in CH$_4$ is usually around 0.73, while that in NO is about 0.43. The lithium loading seems to have no significant effect on the NO partial pressure dependency but it may increase the CH$_4$ dependency somewhat compared to pure MgO. The partial pressure dependency studies were also extended to high CH$_4$ and low NO concentrations over the 4% Li/MgO catalyst and the corresponding NO and CH$_4$ reaction orders determined around P$_{CH4}$ = 33.8 Torr and P$_{NO}$ = 1.6 Torr, respectively, are very close to those at lower CH$_4$ and higher NO concentrations. Kinetic studies of NO reduction by CH$_4$ in the presence of 1.0% O$_2$ were also performed over MgO and Li/MgO samples. The apparent activation energies of 35.0 ± 1.0 kcal/mole are higher than those in the absence of oxygen. Again, no correlation existed between activation energies and lithium loadings. The empirical reaction orders in O$_2$ at 923K determined near P$_{CH4}$ = 3.8 torr and P$_{NO}$ = 15.4 are negative for all catalysts (-0.54) at O$_2$

concentrations below 15 torr, as presented in Table II, but they approach zero order at O$_2$ pressures greater than 15 torr. Further details on NO decomposition in He are provided elsewhere (*14*).

Over La$_2$O$_3$ the apparent activation energy is 24.4 kcal/mol in the absence of O$_2$, which was much lower than that for the Li/MgO catalysts, and it increased to 26.0 kcal/mol in the presence of 1% O$_2$. The reaction orders with respect to CH$_4$ and NO in the absence of O$_2$ were determined by first keeping NO at 15.4 Torr and then by keeping CH$_4$ at 3.8 Torr, and as summarized in Table II, the respective reaction orders are 0.26 and 0.98. The reaction order in O$_2$ was initially examined by keeping NO at 10.8 torr and CH$_4$ at 2.7 Torr at a GHSV of 12,400 h^{-1}, and the apparent reaction order in O$_2$ thus obtained was about 0.4 at low O$_2$ concentrations up to 0.5% but became slightly negative at higher O$_2$ concentrations; however, depletion of CH$_4$ was greater than 70% at O$_2$ concentrations higher than 0.5%. Therefore, the experiment was repeated at a higher GHSV of 49,600 h^{-1} and the results showed behaviour similar to that at low O$_2$ concentrations. In all these experiments where the CH$_4$ pressure was kept at 2.7 Torr in the feed, CH$_4$ conversions were high and since the reaction order in CH$_4$ is ~0.26, the rate of N$_2$ formation should decrease as the CH$_4$ partial pressure decreases. The reaction order in O$_2$ was finally determined at a higher CH$_4$ partial pressure of 25.6 Torr while again keeping the GHSV at 49,600 h^{-1}. Under these conditions CH$_4$ conversions were kept below 12% and the reaction order in O$_2$ thus obtained was 0.5. Additional information about N$_2$O and NO$_2$ reduction and decomposition is given in another paper (*15*).

Discussion

To test our hypothesis that the generation of methyl radicals can enhance the reaction between CH$_4$ and NO$_x$ compounds, we first examined NO reduction by CH$_4$ over MgO and Li/MgO, which is the most characterized catalyst system for methane oxidative coupling. This study showed that Li/MgO catalysts as well as pure MgO are indeed active for NO reduction by CH$_4$ and the activity continuously increased with temperature up to 953K. Both N$_2$ and N$_2$O as nitrogen-containing products were observed, and selectivity to N$_2$ increased with temperature. N$_2$ mass balances showed that little or no NO$_2$ formation occurred. Control experiments demonstrated that there was no reaction of either NO or CH$_4$ in an empty quartz tube reactor with quartz wool inserted. Further experiments in the absence of CH$_4$ showed that there was no direct NO decomposition over these catalysts, which implies that NO reduction can proceed only in the presence of gaseous CH$_4$.

As shown in Table I, varying the Li loadings from 1 to 16% had little effect on the specific activity, but the absence of Li results in an approximate 5-fold decrease. Despite this decline in specific activity, neither the activation energy nor the partial pressure dependencies were strongly dependent on the Li loading, thus allowing the possibility that a similar reaction mechanism exists on all four catalysts. For all catalysts, the apparent E$_{act}$ values were near 29 kcal/mole and the NO pressure dependence was around 0.4. The pressure dependence on CH$_4$ appeared to increase somewhat from 0.6 on pure MgO to 0.9 at a high Li loading of 16%, while the negative dependence on O$_2$ seemed to decrease slightly (Table II). A substantial amount of N$_2$O is produced at temperatures below 900K and the selectivity to N$_2$ is about 60%, giving a N$_2$/N$_2$O ratio of about 1.5; however, at higher temperatures, this ratio increases significantly to 3.0 at 953K with the 4% Li/MgO catalyst, as shown in Figure 1. Over pure MgO, the amount of N$_2$O formed approaches zero at these higher temperatures, thus the presence of Li may facilitate N$_2$O formation. The presence of molecular O$_2$ significantly inhibits N$_2$ formation at higher NO concentration. However, the relative decrease in the rate of N$_2$ formation caused by O$_2$ is much less at higher temperatures.

To further test the hypothesis, we next studied La_2O_3, which is one of the best methane coupling catalysts. This La_2O_3 was precalcined at 1023 K for 10 h to stabilize the surface area, and it exhibited stable performance for NO reduction over a period of 30 h while the BET surface area varied little during reaction. The activity continuously increased with temperature up to 973 K, which was the highest temperature employed in this work. This investigation shows that La_2O_3 is more active for NO reduction to N_2 by CH_4 than MgO and Li-promoted MgO; for example, the specific activity of 1.51 $\mu mol\ N_2/s \cdot m^2$ obtained over La_2O_3 at 773 K in 2.02 %NO and 0.505 %CH_4 is about 13 and 3 times higher than the corresponding values for MgO and 4%Li/MgO, respectively, as shown in Table I. More importantly, oxygen greatly promotes NO reduction by CH_4 over La_2O_3 and the activity increased by a factor of about 5 when 1% O_2 was added. The reaction has a positive partial pressure dependence on O_2 of about 0.5 for La_2O_3, whereas O_2 has an inhibiting effect on NO reduction by CH_4 over Li/MgO catalysts and the reaction order was near -0.5. Furthermore, in contrast to a NO selectivity to N_2 of 60-75% over Li/MgO, N_2 was the only significant nitrogen-containing product observed during NO reduction by CH_4 over La_2O_3 over the entire temperature range; thus the selectivity to N_2 was essentially 100%. When compared to Co/ZSM-5 on a weight basis, this low surface area (ca. 3.4 m^2/g) La_2O_3 is much less active; however, when normalized on a surface area basis, La_2O_3 is similar in activity with no O_2 present and is slightly more active in the presence of O_2, as shown in Table I. As pointed out earlier, Co/ZSM-5 has shown a volcano-like activity with a maximum around 700 K *(7,8)*, whereas the activity of La_2O_3 continuously increased with temperature. As the activity of La_2O_3 is a function of its BET surface area, much higher activities per gram for NO reduction by CH_4 over La_2O_3 should be achievable by supporting it on a stable, high surface area support. This will be done in future studies.

Unlike the MgO and Li/MgO catalysts, La_2O_3 was also quite active for direct NO decomposition; however, the activity was significantly lower than that for NO reduction by CH_4; for example, at 973 K the rates of direct NO decomposition and NO reduction were 1.1×10^{-2} and 3.4×10^{-2} $\mu mol/s \cdot m^2$, respectively, although the apparent activation energies of these two reactions were quite similar, being 23.0 and 24.4 kcal/mol, respectively *(15)*. Consequently the high activity for NO conversion to N_2 on La_2O_3 is due mainly to the presence of CH_4. It is not clear at this point why La_2O_3 is more active than Li/MgO for NO reduction by CH_4 although we currently are associating activity with methyl radical formation. The only previous work on NO reduction involving La_2O_3 is that of Muraki et al. *(16)*, in which the authors were examining Pd dispersed on La_2O_3 as a practical, less expensive alternative to Rh in three-way catalyst systems, and La_2O_3 was found to improve the ability of Pd to catalyze the reduction of NO to N_2 under rich conditions. However, no explanation was given as to how La_2O_3 might promote NO reduction under lean conditions in the absence of the metal. The apparent activation energy for NO reduction by CH_4 over La_2O_3 (24.4 kcal/mol) was noticeably lower than that obtained over Li/MgO (29.0 kcal/mol). Pressure dependencies on CH_4, NO and O_2 over La_2O_3 and Li/MgO were also very different, as summarized in Table II, with the most significant difference being the pressure dependency on O_2, which was 0.50 for La_2O_3 and -0.54 for Li/MgO. Thus the detailed reaction mechanism over La_2O_3 and Li/MgO may be different, although both catalysts can generate methyl radicals under appropriate conditions.

It is important to compare these nonmetallic catalysts with the performance of zeolite catalysts because few catalysts have been able to successfully activate CH_4 for NO_x reduction. One obvious difference between these two catalyst systems is that all the zeolite catalysts have shown a volcano-like (or bending-over) activity dependence on temperature, with the activity maximum falling between 670 and 770K. Li and Armor believe that the decrease in NO conversion at high temperature may be the result of significantly lowered CH_4 concentrations due to combustion *(8)*. However, Petunchi and Hall have suggested that this decrease at higher temperatures for NO reduction by i-C_4H_{10} in the presence of O_2 is due to lowered equilibrium concentrations of nitrogen dioxide or a much larger decrease in the NO_2/NO ratio (17). The rates of NO reduction by CH_4 over both Li/MgO and La_2O_3 catalysts continuously increase up to 953K (for Li/MgO) and 973 K (for La_2O_3) in the absence as well as the presence of O_2. No bending-over behavior in activity of NO conversion to N_2 was observed in the present work up to a NO conversion of 45% over La_2O_3 at 973 K. In the absence of O_2, no NO_2 was detected over any of these samples, thus we cannot confirm that NO_2 may be a stable intermediate in NO reduction over these catalysts. However, our current analytical capabilities make it very difficult to detect NO_2. On the contrary, we suggest that a different mechanism might be applicable for NO reduction by CH_4 over La_2O_3, i.e., N_2O may be an intermediate because it decomposes to N_2 and O_2 so rapidly on La_2O_3 (15). In the presence of O_2, no decrease in NO conversion was observed with temperature even though CH_4 conversions were above 50% at temperatures above 923K for Li/MgO, and 893 K for La_2O_3. When comparing our reaction kinetics with previous work on NO reduction by CH_4, only the results of Li and Armor are available ($7,8$). Over Mn/ZSM-5 and Co/ZSM-5 the CH_4 pressure dependencies were near 0.6 and the NO pressure dependencies were around 0.5, which are similar to those observed over the Li/MgO catalysts. Over La_2O_3, the reaction order in CH_4 (0.26) is lower and that in NO (0.98) is higher than both the zeolite and Li/MgO catalysts. The addition of O_2 up to 0.5% increased NO conversion over Co/ZSM-5 and the apparent activation energy calculated from their results for Co/ZSM-5 is about 12 kcal/mole. A similar result of an oxygen-promoting effect on NO conversion was observed with La_2O_3, but the corresponding activation energy is much higher (26.0 kcal/mol), whereas oxygen inhibits NO conversion over Li/MgO, particularly at low temperature. These comparisons indicate that the reaction mechanism over metal-loaded ZSM-5 and the non-metallic catalysts may be different.

It is well known that methyl radicals can be generated on the surface of Li/MgO catalysts during methane oxidative coupling ($10,11$). Control experiments performed over both 4% Li/MgO and La_2O_3 verify that the catalysts prepared in this study are indeed methane oxidative coupling catalysts, which implies that they can generate methyl radicals under appropriate conditions. Under similar conditions, e.g., at 993 K, a GHSV of 17,400 h^{-1} and a $CH_4:O_2$ ratio of 5:1, CH_4 conversion was 15.6% and selectivity to C_2 hydrocarbons was 42.2% over La_2O_3 (15), which favorably compares with results reported by Lunsford et al., i.e., at 998 K and a $CH_4:O_2$ ratio of about 9:1; their CH_4 conversion was 9.4% and their C_2 selectivity was 46.6% (18). Furthermore, with La_2O_3 both activity and selectivity increased with temperature, and selectivity to C_2 hydrocarbons decreased with increasing O_2 concentration. When the CH_4/O_2 ratio was 1/4 or lower, almost all the CH_4 was oxidized into CO and CO_2, and when N_2O was used instead of O_2, similar results were obtained (15). However, we have not yet been able to verify that methyl radicals are directly involved in NO reduction by CH_4. We could detect no C_2 hydrocarbons under any circumstances during NO reduction by CH_4, even at a CH_4/NO ratio of 5/2 and in the presence of O_2. The similar, though lower, activity of pure MgO in this reaction is consistent with the recent study of Goodman et al. which indicated that oxygen vacancies are involved in CH_4 activation (19). The promotional effect of O_2 on NO reduction by CH_4 over La_2O_3 suggests that methyl radicals are involved in NO reduction in this case. Lunsford et al. have proposed that $[Li^+O^-]$ centers on Li/MgO activate methane by generating methyl radicals ($10,11$). In the present work over La_2O_3, it is obvious that methane activation must occur on other types of active centers. Goodman et al. have utilized surface science techniques along with elevated pressure kinetic measurements to study the conversion of methane to ethane over model MgO and Li/MgO catalysts ($19,20$), and they have suggested that a defect site such as an F-center (an oxygen vacancy containing one or two electrons) in the near-surface region is responsible for methane activation. The reactivity of lattice oxygen in La$_2$O$_3$ at 973 K was investigated by passing only a 1% CH$_4$/He mixture at a flow rate of 10 ml/min through a fully oxidized La$_2$O$_3$ sample; CO$_2$ and some CO were observed as products. The amounts of the products decreased with time and almost no CO$_2$ or CO evolution was detected after 60 min on-stream at 973 K. The amount of lattice oxygen consumed under these conditions was ~0.67% of the total lattice oxygen of the sample, giving a formula of La$_2$O$_{2.98}$ for the La$_2$O$_3$ catalyst after reduction. Presumably the oxygen vacancies are mainly concentrated at the catalyst surface and, with a surface area of 3.5 m^2/g, the total amount of O removal constitutes about half a monolayer. Under reaction conditions, even under a net oxidizing atmosphere, these vacancies should exist to some extent at high temperature and they may well be the active center on La$_2$O$_3$ for CH$_3$· generation, as Goodman and coworkers have proposed.

Assuming methane activation does occur via methyl radicals on either [Li+O-] on Li/MgO or an F-center on La$_2$O$_3$, the absence of any detectable amount of C$_2$ hydrocarbons during NO reduction by CH$_4$ over either catalyst system under any circumstances implies one of several possibilities: (a) once generated, all methyl radicals on the surface react rapidly with NO before they can desorb into the gas phase and couple to form C$_2$H$_6$; (b) methyl radicals do desorb but react rapidly with NO in the gas phase; or (c) any C$_2$ hydrocarbons formed by coupling of methyl radicals react further with NO, as it has been reported that higher hydrocarbons are more reactive with NO than CH$_4$ during NO reduction over zeolite catalysts (3). The gas-phase reaction

$$CH_3· + NO \longrightarrow CH_3NO \qquad (3)$$

has a rate constant of k=1.7x10^{-11} mol/cm^3/s at 298 K, compared to a rate constant of k=3.9x10^{-11} mol/cm^3/s at 973 K for CH$_3$· radical coupling (21), i.e.,

$$CH_3· + CH_3· \longrightarrow C_2H_6 \qquad (4)$$

Considering the enormous difference in the concentrations of NO and CH$_3$·, reaction 3 would be heavily favored over reaction 4; consequently, the probability of producing detectable quantities of C$_2$ hydrocarbons would be very low.

The fact that both CH$_4$ conversion and selectivity to C$_2$ hydrocarbons increase with reaction temperature during oxidative coupling is consistent with the observation that during NO reduction by CH$_4$ the rates of N$_2$ formation continuously increase with temperature. According to Lunsford and co-workers, the concentration of methyl free radicals rises with temperature (10,11). In addition, our recent studies have shown that Sr/La$_2$O$_3$ is an even more active catalyst than La$_2$O$_3$ for NO reduction by CH$_4$, which is consistent with the results of methane coupling over these two catalysts as Sr/La$_2$O$_3$ is more active (22). However, at this time we have not yet been able to prove that methyl radicals are directly involved in these reactions.

Summary

Non-metallic catalysts — MgO, Li/MgO and La$_2$O$_3$ — known to produce methyl radicals during the methane oxidative coupling reaction have been shown to be active for NO reduction by CH$_4$. Li-promoted MgO in the absence of O$_2$ produces N$_2$ and N$_2$O with a (N$_2$/N$_2$O) selectivity below 2 at low temperature but which increases to 3 or more at higher temperatures. Unpromoted MgO is less active but produces almost 100% N$_2$ at high temperatures. La$_2$O$_3$ is more active and selective for NO reduction to N$_2$ by CH$_4$ than MgO and Li/MgO catalysts. The activity of La$_2$O$_3$ continuously increases with temperature, at least up to 973 K, and selectivity for N$_2$ rather than N$_2$O formation is essentially 100%. No bending-over in activity of N_2 formation was observed up to NO conversion of 45%. When rates are compared to a Co/ZSM-5 catalyst on a square meter basis, the specific activity of La_2O_3 at 773 K is comparable to that of the zeolite. The most significant difference in kinetics between La_2O_3 and Li/MgO catalysts is that the presence of O_2 in the feed markedly enhances the rate of NO reduction by CH_4 over La_2O_3 with reaction order of 0.5, while it has an inhibiting effect on NO reduction over Li/MgO with a reaction order on O_2 near -0.5.

Acknowledgments

This study was supported by the National Science Foundation under Grant No. CTS-9211552 and by the Florida Power and Light Company.

Literature Cited

1. Armor, J.N., *Appl. Catal.,* **1992,** *B1*, 221.
2. Truex, T.J., Searles, R.A., and Sun, D.C., *Plantinum Metals Rev.,* **1992,** *36*, 2.
3. Iwamoto, M., and Hamada, H. *Catal. Today,* **1991,** *10*, 57.
4. Burch, R., and Millington, P. J., *Appl. Catal.,* **1993,** *B2*, 101.
5. Li, Y., and Hall, W.K., *J. Phys. Chem.,* **1990,** *94*, 6145.
6. Ansell, G.P., Diwell, A.F., Golunski, S.E., Hayes, J.W., Rajaram, R.R., Truex, T.J., and Walker, A.P., *Appl. Catal.,* **1993,** *B2*, 81.
7. Li, Y., and Armor, J.N., *Appl. Catal.,* **1992,** *B1*, L31.
8. Li, Y., and Armor, J.N., *Appl. Catal.,* **1993,** *B2*, 239.
9. Li, Y., and Armor, J.N., *US Patent,* **1992,** 5, 149, 512.
10. Ito, T., and Lunsford, J.H., *Nature,* 1985, *314*, 721.
11. Ito, T., Wang, J.-X., Lin, C.-H., and Lunsford, J.H., *J. Am. Chem. Soc.,* **1985,** *107*, 5062.
12. Xu, M., Shi, C., Yang, X., Rosynek, M.P., and Lunsford, J.H., *J. Phys. Chem.,* 1992, *96*, 6395.
13. Palmer, M., and Vannice, M.A., *J. Chem. Technol. and Biotech.,* **1980,** *30*, 205.
14. Zhang, X., Walters, A.B., and Vannice, M.A., *J. Catal.,* **1994,** *146*, 568.
15. Zhang, X, Walters, A.B., and Vannice, M.A., *Appl. Catal.,* **1994,** (in press).
16. Muraki, H., Yokota, K., and Fujitani, Y., *Appl. Catal.,* **1989,** *48*, 93.
17. Petunchi, J.O. and Hall, W.K., *Appl. Catal.,* **1993,** *B2*, L17.
18. Lin, H., Campbell, K.D., Wang, J.X., and Lunsford, J.H., *J. Phys. Chem.,* **1986,** *90*, 534.
19. Wu, M.-C., Truong, C.M., Coulter, K., and Goodman, D.W., *J. Catal.,* **1993,** *140*, 344.
20. Coulter, K., and Goodman, D.W., *Catal. Lett.,* **1993,** *20*, 169.
21. Lunsford, J.H., personal communication.
22. Zhang, X, Walters, A.B., and Vannice, M.A., (in preparation).

RECEIVED November 22, 1994

Chapter 7

Low-Temperature Selective Catalytic Reduction of NO by Hydrocarbons in the Presence of O_2 and H_2O

Benjamin W.-L. Jang[1], James J. Spivey[1], Mayfair C. Kung[2], Barry Yang[2], and Harold H. Kung[2]

[1]Center for Process Research, Research Triangle Institute, Research Triangle Park, NC 27709
[2]Department of Chemical Engineering, Northwestern University, Evanston, IL 60208

The low-temperature selective catalytic reduction of NO over an active carbon-supported catalyst was studied in order to develop catalysts that are active at temperatures around 150 °C as part of a low-temperature process for NO control in flue gases from coal-fired power plants. Reductants studied include acetone, isopropanol, isobutanol, ethyl ether, ethanol, propene, and methanol. Acetone is the most active reductant for the selective reduction of NO over the 5%Cu-2%Ag/C catalyst tested. At 150 °C, 35% NO conversion was obtained in the presence of 4% O_2 and 8% H_2O at 3,000 h^{-1} space velocity after 5 h on stream. There is some decrease in NO and hydrocarbon conversion with time on stream. It is believed that the oxidation of acetone minimizes the oxidation of carbon in the presence of O_2 and promotes the selective reduction of NO.

The selective catalytic reduction of nitric oxide in the presence of oxygen and water vapor is one of the key research areas driven by the recent Clean Air Act Amendments. A wide range of NO_x control processes for electric utilities are under investigation. Many are approaching commercial feasibility and will compete with the current processes such as selective catalytic reduction (SCR). Although SCR using NH_3 is effective for NO reduction, the process suffers from many disadvantages, e.g., short catalyst life, safety problems associated with ammonia storage and transportation, ammonia slip, and plugging caused by ammonium sulfate salt formation. In addition, current SCR processes must be located at places within the flue gas treatment process where temperatures are near 400 °C. Retrofitting SCR into these locations in the plant is costly because space and access in many power plants is extremely limited (1). Therefore, there is a need to develop NO reduction technologies that can be located downstream of the particulate control device, near the stack, where temperatures are around 150 °C. These temperatures are also

0097–6156/95/0587–0083$12.00/0

typical of industrial boiler flue gases near the stack, suggesting another application for this technology.

Although NH$_3$ is active and selective for NO reduction at low temperatures (2), very little work has been reported using hydrocarbon reductants at temperatures below 200 °C. Hydrocarbons are less expensive and safer to handle than NH$_3$, and a hydrocarbon-based process represents a reasonable alternative to the ammonia-based SCR commercial processes for NO reduction. Recently, it was found that NO can be selectively reduced by hydrocarbons in excess oxygen over many different catalysts (3-8). For example, Cu-ZSM-5 has been shown to be an excellent catalyst for NO reduction using hydrocarbons (except methane) in excess oxygen (3,4,7). However, the activity of the catalyst declines rapidly at temperatures below 300 °C (9). The most active catalyst for low-temperature NO reduction using hydrocarbons has been reported by Kung et al. (10). At 265 °C, 90% NO conversion was reported using propene at a 1/1 NO/propene stoichiometry over a 13.6% Cu-Ga$_2$O$_3$ catalyst in the presence of oxygen at a W/F of 0.017 g-min/cc. However, there was no water vapor in the simulated exhaust stream in these tests. Later tests in the presence of 5% water vapor and 4% O$_2$ found that isobutanol is a more active reductant for NO and that an La-promoted Cu-ZrO$_2$ catalyst was the most active of those tested, achieving 65% NO conversion at 275 °C, and W/F of 0.020 g-min/cc (11). At a lower water vapor concentration of 2%, 83% conversion of NO to N$_2$ was achieved on this catalyst using isobutanol in the presence of 4% O$_2$ at a space velocity of 3,000 h^{-1} and a temperature of 200 °C (12). This is believed to be the highest conversion reported to date using a hydrocarbon reductant.

However, there is a need to extend the operating temperature for the selective reduction of NO even lower. The low-temperature activity of carbon-based catalysts for the SCR reaction (2) suggests that they could be modified for the selective reduction of NO using hydrocarbons. The development of a high activity carbon-based catalyst for the selective reduction of NO at temperatures of 150 °C and below has recently been reported by Gangwal et al. (13), although NH$_3$ was used as the reductant. A carbon-supported noble metal catalyst achieved 90% NO reduction at 158 °C and 3,400 h^{-1} using NH$_3$. Recent work by Yamashita and Tomita (14) showed that the addition of Cu to an activated carbon support promoted the direct reduction of NO by carbon in the presence of O$_2$ (2 NO + C → CO$_2$ + N$_2$), while minimizing the oxidation of the carbon itself. Even though this study was carried out at temperatures above 300 °C (without added hydrocarbons) and had as its object the consumption of the carbon to reduce NO, it suggests the role of Cu/C in activating the NO molecule in the presence of O$_2$. Miyadera (15) showed that supported Ag can be used to promote the selective reduction of NO in the presence of H$_2$O and O$_2$ using a range of hydrocarbon reductants such as acetone and C$_1$-C$_3$ alcohols. NO conversion was above 90% at 275 °C. However, conversion decreased rapidly below about 250 °C, and carbon was not among the supports investigated.

Therefore, this study seeks to extend the selective reduction of NO to temperatures near 150 °C in the presence of realistic concentrations of H$_2$O and O$_2$ using hydrocarbon reductants on a Cu-Ag/C catalyst.

Experimental

Catalyst Preparation. PCB active carbon from calgon (surface area ~1,000 m^2/g) was used in this study. All the granules were broken into 16/30 mesh particles followed by drying at 120 °C overnight. Ultrapure copper nitrate and silver nitrate, used for the preparation of catalysts, were purchased from Johnson Matthey. The 5%Cu-2%Ag supported catalysts were prepared by pore volume impregnation. The impregnated catalysts were first dried under vacuum at room temperature for 2 h, then at 60 °C for 2 h and 80 °C overnight. All catalysts were treated with 150 cc/min of helium at 400 °C for 1 hour before reaction.

Catalytic Tests. The schematic of the experimental system used in this study is shown in Figure 1. The system consisted of a convective oven (Varian 3700 GC oven) to provide temperature control both for the reaction studies at temperatures of 100 to 250 °C and for the catalyst pretreatment at 400 °C. A Pyrex reactor tube of 1/4 in. on both ends and 3 in. × 3/4 in. on the body was used. Stainless-steel tubing (1/4 in.) was used to connect the Pyrex reactor on both ends to the rest of the system. Two 1/16 in. thermocouples were positioned in the reactor, one at the inlet and one at the outlet. The catalyst bed was positioned in the center between glass wool supports. An extra stainless-steel coil before the inlet reactor was provided to increase the length of the preheat zone in the oven. For all experiments 400 cc/min of total flow and 7 g of catalyst were used (W/F = 0.0175 g-min-cm^{-3}, space velocity 3,000 h^{-1}). The desired gas mixture and steam were produced by four mass flow controllers and a syringe pump. The gas cylinders used were He, certified O_2/He, certified NO/He, and certified C_3H_6/He obtained from a commercial vendor. The reactant gas mixture contains 0.1% NO, 0.13% acetone, 4% O_2, and 8% H_2O if not otherwise specified. All tubing was heated by heating tapes to prevent any condensation. A chemiluminescent NO_x analyzer (Thermo Electric Model 10S) was used to monitor the inlet and outlet NO_x concentration continuously. A 486 personal computer with HP Chem station was used to control the gas chromatograph (GC) injection, separation, and integration of reaction products.

The reaction gases were analyzed using an online Hewlett Packard Model 5890 GC equipped with a thermal conductivity detector (TCD) and two columns in series. The first column, 9 ft × 1/8 in. o.d. stainless-steel column packed with 60/80 mesh Hayesep R, was used to separate hydrocarbons, CO_2 and N_2O from O_2, N_2, and CO. The second column, a 6 ft × 1/8 in. o.d. stainless-steel column packed with 45/60 mesh molecular sieve 5A, was used to separate O_2, N_2, and CO. After injection of a sample into the Hayesep R column at 35 °C, O_2, N_2, and CO were allowed to elute into the molecular sieve column. The molecular sieve column was then bypassed and the CO_2 and N_2O were separated and eluted from the Hayesep R column and into the TCD. The molecular sieve column was then placed back on-line and the O_2, N_2, and CO were separated and detected by TCD while oven temperature increased 20 °C/min from 35 to 100 °C. Finally, the molecular sieve column was bypassed again and hydrocarbons were separated and eluted from the Hayesep R column when oven temperature increased 35 °C/min to 220 °C. The reaction results

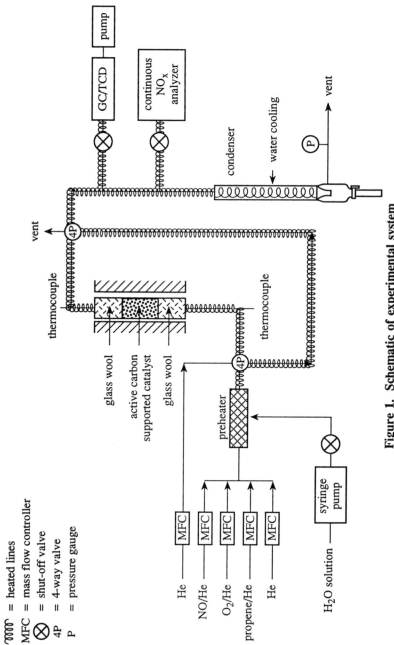

Figure 1. Schematic of experimental system

are described in terms of NO conversion, N_2 selectivity N_2O selectivity, and hydrocarbon conversion:

$$\%NO \text{ conversion} = (NO_{in} - NO_{out}) \cdot 100/NO_{in}$$

$$\%N_2 \text{ selectivity} = N_2 \cdot 100/(N_2 + N_2O)$$

$$\%N_2O \text{ selectivity} = 100 - \%N_2 \text{ selectivity}$$

$$\%HC \text{ conversion} = (HC_{in} - HC_{out}) \cdot 100/HC_{in}$$

Results and Discussion

Mass Balances. Mass balances on both carbon and nitrogen atoms were in the range of 70 to 80% for all tests. In the case of nitrogen, this is apparently not due to the formation of NO_2, since the extremely high moisture content of the gas would be expected to lead to the adsorption of nitric acid on the walls of the system downstream of the reactor. However, there was no difference in the results with and without water vapor, suggesting that the incomplete nitrogen balance is not due to the adsorption of NO_2. In the case of carbon atoms, there may have been some adsorption in the micropores of the carbon, especially for water-soluble reductants such as acetone. This may account in part for the transient behavior seen for all reductants. For carbon atoms, the mass balance improved slightly with time on stream, but never exceeded about 80% even for long reaction times, so that adsorption alone does not account for the shortfall in the mass balance. The calibration for the hydrocarbons and carbon-containing reaction products did not change during the tests, thus analytical error can be ruled out. Further efforts are under way to determine if there are any undetected carbon-containing products.

HCN Formation. Recently, it has been shown that HCN can be formed in the reduction of NO with propene and ethene (*16*), but the concentrations did not exceed about 30 ppm in that study, which would not close the mass balance on either nitrogen or carbon atoms. In addition, HCN was formed on an acidic support, ZSM-5, which is unlike the catalysts used here. Also, HCN formation was maximized at temperatures between 250 and 320 °C, depending on the reductant. HCN formation was negligible at temperatures below about 200 °C.

Transient Behavior. Transient behavior was observed in all tests. The NO conversion decreased with time on stream from initially high values to much lower ones even for the most active reductants. The reasons for this are unclear, but do not appear to be due to NO adsorption, as discussed below. Because the catalyst was calcined in He rather than in air, it may be that the oxidation state of the Cu, which has been shown to be a key to the selective reduction of NO (for example, see reference 17), may change during the initial part of the reaction.

Propene and Acetone Results.

Propene Versus Acetone. The NO conversion over 5%Cu-2%Ag/C with acetone and propene as reductants in the presence of O$_2$ and H$_2$O is shown in Figure 2, along with the results in the absence of any hydrocarbon. The NO conversion with acetone as reductant became essentially constant at around 35% after 5 h on stream. On the other hand, the NO conversion with propene as the reductant decreased to <10% after 3 h on stream. Acetone is a more active reductant than propene for NO selective reduction in the presence of O$_2$ and H$_2$O over the 5%Cu-2%Ag/C catalyst at 150 °C.

Table I compares the conversion of NO and the two hydrocarbons, as well as the selectivity to N$_2$ (all calculated at steady state), to the results without an added hydrocarbon reductant. The conversion of acetone is greater than can be accounted for by the reduction of NO according to the following reactions:

$$(CH_3)_2CO + 2NO + 3O_2 \rightarrow 3CO_2 + N_2 + 3H_2O \qquad (1)$$

$$(CH_3)_2CO + 8NO \rightarrow 3CO_2 + 4N_2 + 3H_2O \qquad (2)$$

The ratio of acetone reacted to NO reacted at steady state (from Figure 2) is 2.2. This compares to values of 0.5 (reaction [1]) and 0.125 (reaction[2]) that would be predicted from the NO reduction reactions. This suggests that the oxidation of

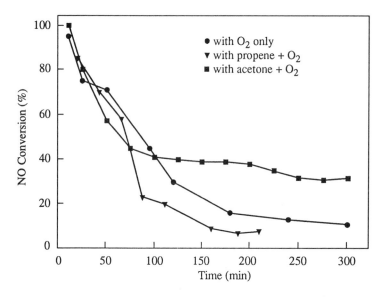

Reaction conditions: 1,000 ppm NO, 4% O$_2$, 8% H$_2$O, 3,000 h^{-1}, 150 °C.

Figure 2. NO conversion versus time over 5%Cu-2%Ag/C for the reaction with acetone+O$_2$, propene+O$_2$, and O$_2$ only

Table I. Comparison of NO Reduction with Acetone and Propene[a]

Hydrocarbon reductant	Hydrocarbon conversion (%)	NO conversion	N_2 selectivity
acetone[b]	69	35	>99
propene[c]	<1	7[d]	>99
none	—	11	>99

[a] Reaction conditions: 1,000 ppm NO, 4% O_2, 8% H_2O, 3,000 h^{-1}, 150 °C, plus reductant (see notes b and c below). Data shown are for steady state from Figure 2.
[b] 1,300 ppm.
[c] 1,100 ppm.
[d] After 3 hours on stream.

acetone, either to CO_2+H_2O or to partial oxidation products (which were not detected by our analytical system), is significant even at these low temperatures.

Propene oxidation, as measured by the difference between the inlet and outlet concentrations of propene, was negligible over 5%Cu-2%Ag/C in the presence of O_2 and H_2O at 150 °C (Figure 2 and Table I). The conversion of NO with propene as reductant is actually lower than without propene. Just as with added reductants, NO conversion decreased rapidly to <20% after 3 h in the presence of oxygen only. The reduction of NO in the absence of an added gas-phase reductant can be attributed to the reaction of the carbon with NO:

$$C + 2NO \rightarrow CO_2 + N_2 \qquad (3)$$

This reaction is known to be promoted by Cu on carbon in the presence of O_2 at 300 °C (*14,18*) and is apparently significant on this catalyst at 150 °C. The rate of production of CO_2 without propene is higher than with propene, which further supports this hypothesis. The lower NO conversion in the presence of propene suggests a kinetic inhibition of the C+NO reaction by propene in the presence of O_2. Propene probably competes with NO for reactive sites on the carbon surface (not necessarily metal sites), which in turn decreases the availability of carbon for the reduction of NO via reaction (3).

Effect of Oxygen. The effect of oxygen on the reduction of NO by acetone is shown in Figure 3. The NO conversion was low in the beginning and decreased to ~10% after 20 min on stream for the reaction with acetone only. In the presence of oxygen, however, NO conversion was high and slowly decreased to a steady level. This shows that the reaction of NO directly with acetone (reaction [2]) is less significant than the selective reduction reaction (3) at these conditions, accounting for only about 10% NO conversion out of a total measured NO conversion of 40% after 100 min on stream. Although not conclusive, the difference in NO conversion

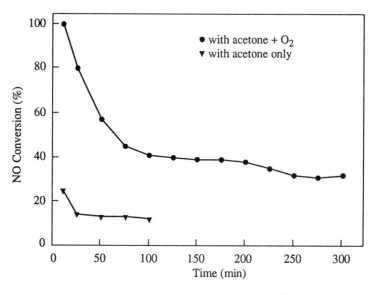

Reaction conditions: 1,000 ppm NO, 8% H$_2$O, 3,000 h^{-1}, 150 °C.

**Figure 3. NO conversion versus time over
5%Cu-2%Ag/C with acetone+O$_2$ and acetone only**

with and without oxygen suggests that physical adsorption of NO did not contribute to the initially high NO conversion in the acetone+O$_2$ case. If it did, the NO conversion with time would be similar whether oxygen was present or not. Another possibility is that NO$_2$ is formed by oxidation of NO in the acetone+O$_2$ case. At the high water vapor concentrations of these tests, NO$_2$ would then adsorb as nitric acid, which could cause the high NO conversion at the beginning of the acetone+O$_2$ test (Figure 3). However, the results in Figure 2 show a large difference in activity between the reaction with propene+O$_2$ and the reaction with acetone+O$_2$, suggesting that there is little NO$_2$ adsorption or nitric acid formation after the first hour. However, the contribution of NO$_2$ adsorption or subsequent reaction with H$_2$O to form nitric acid cannot be ruled out for high NO conversion in the first hour of reaction. The role of NO$_2$ in the reaction of NO selective reduction by hydrocarbons will be further investigated.

Effect of Metals on Carbon Activity. Carbon alone is known as a catalyst for the selective reduction of NO with ammonia (2). Its activity for selective reduction with hydrocarbons has not been well studied. Figure 4 shows the NO conversion versus time for the carbon alone and for the 5%Cu-2%Ag/C catalyst. The enhancing effect of Cu+Ag on the activity of active carbon for NO conversion can be clearly seen with acetone as a reductant. The NO conversion decreased rapidly on active carbon alone (without Cu+Ag) with acetone in the presence of O$_2$ and H$_2$O. NO conversion was negligible after 1 h of reaction. On the other hand, the NO

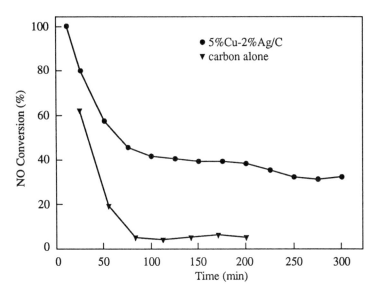

Reaction conditions: 1,000 ppm NO, 8% H_2O, 3,000 h^{-1}, 150 °C.

**Figure 4. NO conversion versus time over
5%Cu-2%Ag/C and carbon only**

conversion over 5%Cu-2%Ag/C slowly decreased to 35% after 5 h. The decrease
in NO conversion coincides directly with a corresponding decrease in the conversion
of acetone as shown in Figure 5. This correlation between the conversion of acetone
and NO (Figure 5), with the results showing that acetone in the absence of oxygen
is not an effective reductant for NO (Figure 3), suggests that the function of Cu+Ag
is to promote the oxidation of acetone for the selective reduction of NO. The carbon
balance for the tests shown in Figure 5 was about 80% after 5 h of reaction. This
indicates that some products of the partial oxidation of acetone may not be detected
by our current system.

Effect of Water Vapor. The effect of H_2O on NO selective reduction over
carbon-supported catalysts was also studied. The literature shows an inhibiting effect
from the addition of H_2O on NO reduction over a number of catalysts, e.g., Cu-ZrO_2
(*11,12*). In contrast, results on the 5%Cu-2%Ag/C (Figure 6) show that H_2O has no
effect on NO reduction. The conversion of NO with 8% H_2O or without H_2O is
essentially the same. This also rules out the possibility that NO_2 further reacts with
H_2O to form nitric acid under the reaction conditions tested, which might contribute
to the conversion of NO.

Selectivity to N_2. Possible nitrogen-containing reaction products in the exit
gas (besides NO) include NO_2, N_2, and N_2O. With one exception, all experiments
over 5%Cu-2%Ag/active carbon in the presence of O_2 showed nearly 100% N_2

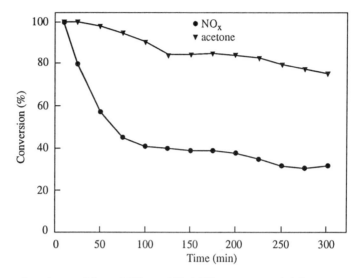

Reaction conditions: 1,000 ppm NO, 1,300 ppm acetone, 4% O$_2$,
8% H$_2$O, 3,000 h^{-1}, 150 °C.

Figure 5. Conversion versus time over 5%Cu-2%Ag/C

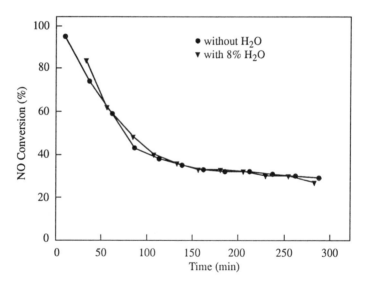

Reaction conditions: 1,000 ppm NO, 1,300 ppm acetone, 4% O$_2$,
3,000 h^{-1}, 150 °C.

Figure 6. NO conversion versus time over 5%Cu-2%Ag/C

selectivity, i.e., no N_2O was formed. The only exception is the NO reduction with acetone in the absence of O_2, which showed >80% N_2O selectivity, suggesting that the NO reduction pathway in the presence of O_2 is different from the pathway in the absence of O_2. The formation of N_2O is probably from the direct reaction between NO and acetone or from the disproportionation of NO (*19*). The selective reduction of NO with propene over Pt/carbon catalysts showed low N_2 selectivity, between 31 and 67%, with higher selectivities at lower temperatures (*12*). In this same study, Cu/C showed essentially complete N_2 selectivity at the same conditions. The high selectivity of the 5%Cu-2%Ag/C catalyst may be related to low catalyst activity for the activation of molecular oxygen, as observed on some solid acid catalysts (*7*). Further research is under way to investigate the function of Cu for NO selective reduction.

As mentioned earlier, the NO selective reduction using acetone in the absence of O_2 showed low activity and low N_2 selectivity. It is clear that gas phase O_2 is required for both high NO conversion (Figure 3) and high N_2 selectivity. It has been suggested that a partially oxidized hydrocarbon intermediate is the active species to reduce NO in the reaction (*20*) and that oxygenated hydrocarbons should be effective for NO selective reduction in the absence of O_2 (*8*). However, this is not the case in our study. The conversion of NO is negligible with acetone in the absence of O_2 (Figure 3). On the other hand, acetone is very effective for NO selective reduction in the presence of O_2. Oxygen is necessary for the selective reduction of NO, at least with acetone. Although it may well be a partially oxidized intermediate on the catalyst surface that selectively reduces NO, these surface intermediates do not appear to be simply physisorbed hydrocarbons present in the inlet gas, e.g., a ketone in this case. Rather, the oxygenated hydrocarbon reductant in the inlet gas is oxidized on the catalyst surface and subsequently reacts with NO or NO_2 (*20*).

Other Oxygenated Hydrocarbons Reductants. Other oxygenated hydrocarbons— 2-propanol, ethanol, methanol, iso-butanol, ethyl ether—were also tested. The inlet carbon atom concentration of these hydrocarbons was calculated to be equal to that of acetone in the tests reported above, e.g., equivalent to 1,300 ppm acetone. The NO conversion at steady state for each reductant is shown in Table II. All reductants showed 100% selectivity to N_2. Since the carbon concentration of all reductants was the same, the activity of the reductants can be compared by comparing the NO conversion. The only reductant with activity close to acetone is 2-propanol with 31% conversion. Others showed much less activity than acetone. Specifically, methanol showed negligible NO reduction activity. It is speculated that the NO selective reduction activity is closely related to the ability of hydrocarbons to form oxygenated surface intermediates at these reaction conditions. This is being investigated further.

Conclusions

Acetone is the most active reductant for NO selective reduction over 5%Cu-2%Ag/C among the hydrocarbons tested. It is believed that the partial oxidation of acetone contributes to the reduction of NO in the presence of O_2 and minimizes the reaction deactivation caused by carbon consumption with O_2 at the same time. Acetone,

Table II. NO Selective Reduction by Hydrocarbons Over
5%Cu-2%Ag/C (1,000 ppm NO, 4% O$_2$, and 8% H$_2$O,
3,000 h^{-1}, 150 °C)[a]

Reductant	NO Conversion (%)	N$_2$ Selectivity (%)
0.13% acetone	35	>99
0.13% 2-propanol	31	>99
0.10% isobutanol	25	>99
0.08% ethyl ether[b]	18	>99
0.20% ethanol	<10	>99
0.40% methanol	negligible	>99

[a] All catalysts subjected to He treatment at 400 °C for 1 h before reaction.
The conversion and selectivity was taken after 5 h on stream.
[b] The concentration of ethyl ether was limited by its solubility in water.

oxygen, and active carbon-supported catalysts are all required to maintain high NO conversion and high N$_2$ selectivity at these low temperatures. Although 35% NO conversion is considered to be high at 150 °C in the presence of O$_2$ and H$_2$O, further improvement is needed for the process to be competitive with other processes. One approach to improve the activity for NO selective reduction to N$_2$ is to minimize the deactivation observed in all the tests, possibly by stabilizing the Cu.

Acknowledgments

This work was prepared with the support, in part by grants made possible by the Illinois Department of Energy and Natural Resources through its Coal Development Board and Illinois Clean Coal Institute, and by the U.S. Department of Energy (Contract 93-1-3-1A-3P). However, any opinions, findings, conclusions, or recommendations expressed herein are those of the author(s) and do not necessarily reflect the views of IDENR, ICCI, and the DOE.

Literature Cited

1.   Wood, S. C. *Chem. Eng. Prog.* **1994**, *37*.
2.   Spivey, J. J. *Annual Rep. Prog. Chem., Part C*, Royal Society of Chemistry: Cambridge, UK, 1994.
3.   Petunchi, J. O.; Gustave, S.; Hall, W. K.   *Appl. Catal. B: Env.* **1993**, *2*, 303-321.
4.   Gopalakrishnan,   R.;   Stafford, P. R.;   Davidson, J. E.;   Hecker, W. C.; Bartholomew, K. H. *Appl. Catal. B: Env.* **1993**, *2*, 165-182.

5.   Miyadera, T.; Yoshida, K. *Chem. Lett.* **1993**, 1483-1486.
6.   Hamada, H.; Kintaichi, Y.; Sasaki, M.; Ito, T.; Tabata, M. *Appl. Catal.* **1991**, *75*, L1-L8.
7.   Hamada, H.; Kintaichi, Y.; Yoshinari, T.; Tabata, M.; Sasaki, M.; Ito, T. *Catal. Today.* **1993**, *17*, 111-120.
8.   Montreuil, C. N.; Shelef, M. *Appl. Catal. B: Env.* **1992**, *1*, L1-L8.
9.   Iwamoto, M.; Mizuno, N.; Yahiro, H. In *Proc. 10th Intern. Cong. Catal.* Budapest. 1992, p.213.
10. Kung, M. C.; Bethke, K. A.; Kung, H. H. *Div. Petro. Chem. Preprints,* 207th ACS National Meeting: San Diego, CA, March 13-18, 1994, pp. 154-158.
11. Kung, H. H.; Kung, M. C.; Yang, B.; Spivey, J. J.; Jang, B. W. -L. Technical report for March-May to the Illinois Clean Coal Institute under ICCI contract 93-1/3.1A-3P: Urbana, IL. 1994.
12. Kung, H. H.; Kung, M. C.; Yang, B.; Spivey, J. J.; Jang, B. W. -L. Technical report for Dec.-Feb. to the Illinois Clean Coal Institute under ICCI contract 93-1/3.1A-3P: Urbana, IL. 1994.
13. Gangwal, S. K.; Howe, G. B.; McMichael, W. J.; Spivey, J. J. "A Novel Carbon-Based Process for Flue-Gas Cleanup"; final report to the U.S. Department of Energy: Morgantown, WV, 1993.
14. Yamashita, H.; Tomita, A. *Energy & Fuel.* **1993**, *7*, 85-89.
15. Miyadera, T. *Appl. Catal. B: Env.* **1993**, *2*, 199-295.
16. Radtke, F.; Koeppel, R. A.; Baiker, A. *Appl. Catal. A: Gen.* **1994**, *107*, L125-L132.
17. Pentuchi, J. O.; Hall, W. K. *Appl. Catal B: Env.* **1994**, *3*, 239-257.
18. Inui, T.; Otowa, T.; Takegami, Y. *Ind. Eng. Chem. Prod. Res. Dev.* **1982**, *21*, 56-59.
19. Li, Y.; Armor, J. N. *Appl. Catal.* **1991**, *76*, L1.
20. Sasaki, M.; Hamada, H.; Kintaichi, Y.; Ito, T. *Catal. Lett.* **1992**, *15*, 297-304.

RECEIVED February 8, 1995

Chapter 8

Catalytic Reduction of NO with Propene over Coprecipitated $CuZrO_2$ and $CuGa_2O_3$

Mayfair C. Kung, Kathleen Bethke, David Alt, Barry Yang, and Harold H. Kung

Ipatieff Laboratory, Center for Catalysis and Surface Science, Northwestern University, Evanston, IL 60208–3000

Copper-zirconium oxide and copper-gallium oxide prepared by coprecipitation and gelation were found to be active catalysts for the selective reduction of NO to N_2 with propene in the presence of a large excess of O_2. The most active catalysts were comparable in activity to the Cu-ZSM-5 catalyst. The dispersion of the Cu ions and the efficiency for propene to react with NO instead of O_2 increased with decreasing Cu content in the catalyst, although the activity decreased. The apparent activation energy for propene oxidation was 17 ± 3 kcal/mol for the 2.1 wt.% and 20 ± 3 kcal/mol for the 33 wt.% Cu-ZrO_2. The apparent activation energy for NO conversion to N_2 was 10 ± 3 kcal/mol for the 2.1 wt.% and 14 ± 3 kcal/mol for the 33 wt.% Cu-ZrO_2 catalysts. The activity of the matrix oxides (ZrO_2 and Ga_2O_3) in reducing NO_2 with propene was much higher than in reducing NO, and the difference between the two matrix oxides correlated with the different response of NO reduction activity on the Cu-Ga_2O_3 and Cu-ZrO_2 to increasing O_2 partial pressure.

Nitric oxide (NO) is an atmospheric pollutant, and its efficient removal from engine exhaust or from the flue gas of boilers and power generation plants is important for the environment. In recent years, much attention in catalytic NO removal has centered on its reduction with environmentally acceptable hydrocarbon molecules in the presence of a large excess of oxygen (lean condition). The lean condition applies to flue gas from power plants, which typically contain a few percent oxygen, and the exhaust of lean burn automobile engines which are under development. Recently, Cu-ZSM-5 catalysts were found to have a high activity for NO reduction under the lean condition *(1-3)*. To date, they are still among the most active catalysts known *(4)*. Unfortunately, Cu-ZSM-5 catalysts suffer from the lack of hydrothermal stability *(5)*, such that they are not suitable for the extended usage required for automobile exhaust converters.

0097–6156/95/0587–0096$12.00/0
© 1995 American Chemical Society

One strategy for the search of catalysts to replace Cu-ZSM-5 would be to assume that the unusually high activity of Cu-ZSM-5 is due to the high dispersion of the Cu ions which are stabilized by the framework charge of the zeolite, and to mimic this situation by dispersing Cu ions in other matrices. Using this approach, we have investigated the NO reduction reaction on mixed oxides of Cu and Zr prepared by the sol gel method *(6,7)*. It was found that such catalysts are quite active in the reduction of NO with propene. In fact, Ni and Co ions well dispersed in ZrO_2 were found also to be effective NO reduction catalysts. Walker et al. *(8)* have independently reported the high activities of $Cu-ZrO_2$ catalysts, and Montreuil et al. *(9)* have reported improved performance of Cu-ZSM-5 catalyst in the presence of $Cu-ZrO_2$. These interesting results prompted further research to characterize $Cu-ZrO_2$ and other similar catalysts. This paper reports the results of the investigation of $Cu-ZrO_2$ and $Cu-Ga_2O_3$ catalysts for the reduction of NO by propene.

Experimental

The catalysts were prepared by coprecipitation and gelation of the nitrate salts of Cu (Fisher Scientific, ACS reagent grade) and Zr (Aldrich, 99.99% $ZrO(NO_3)_2$) with urea at about 373 K using the methods of Amenomiya et al. *(10)*. For catalysts containing Ga, the corresponding nitrate salts were used. After precipitation, the solid was boiled vigorously for 3-4 h to remove excess urea. Then the precipitate was suction filtered, washed with water, filtered again, and dried in air at 373 K. The dried powder was ground and heated in flowing air at a rate of 1 K/min to 623 K and then at that temperature for 3 h. Afterwards, the solid was quenched cooled to room temperature. Before reaction, the catalyst was treated in a flow of 1000 ppm C_3H_6, and 1% O_2 in He at 573 K. This procedure was used to eliminate the residue nitrate left from the preparation procedure and the treatment was terminated when N_2 production from nitrate reaction with C_3H_6 ceased. Cu-ZSM-5, with a Cu loading of 3.2% and a Si/Al ratio of 70, was supplied by GM Corporation, and was the same as the one used by Cho *(11)*.

Reaction tests were conducted using a fused silica U-tube reactor. In most experiments, 1 g of catalyst was tested with a feed of 1000 ppm NO, 1000 ppm propene, and 1% O_2 in He. N_2, O_2, and CO in the gaseous product were analyzed with a 1.5 ft. Carbosphere column linked in series with a 1.5 ft. molecular sieve 5A column at room temperature. CO_2 and hydrocarbons were analyzed with a 10 ft. Porapak Q column at 403 K. N_2O was analyzed by the same Porapak column at 303 K. For activation energy studies, 25 mg of $Cu-ZrO_2$ with particles of 170+ mesh size was used and the O_2 concentration in the feed was 2%. NO conversion to N_2 was kept below 10% and the propene conversion was kept below 30% to ensure that the reactor was being operated in the differential regime. For the data reported, % NO conversion to N_2 and % NO competitiveness are defined as:

% NO conversion $\quad = 2*N_2*100/NO_{in}$ \hfill (1)

% NO competitiveness $\quad = 2*N_2*100/(n*\text{reductant consumed})$

$\qquad\qquad\qquad\qquad\ = 2*N_2*100/(9*\text{propene consumed})$ \hfill (2)

In these equations, N$_2$ is the molar rate of N$_2$ produced, NO$_{in}$ is the molar feed rate of NO, reductant consumed is the molar rate of consumption of propene, and n is the number of O atoms required to completely convert the reductant (propene) into CO$_2$ and H$_2$O, which is 9 for the case of propene. The % NO competitiveness measures the effectiveness (selectivity) of the hydrocarbon reductant to convert NO to N$_2$ versus reacting with O$_2$. With this definition, reactions leading to the formation of N$_2$O would not contribute to the % NO competitiveness. The term "% selectivity" is not used to avoid confusion with the term "selective reduction" of NO to N$_2$.

X-ray photoelectron spectroscopy (XPS) analyses were performed in a VG Scientific spectrometer and sensitivity factors of 4.0 for Cu 2p$_{3/2}$ and 1.1 for Zr 3d$_{5/2}$ were used. Electron paramagnetic resonance (EPR) analyses were performed with a modified Varian E-4 X-band spectrometer. Temperature programmed reduction (TPR) was performed in the tubular flow reactor using 5% H$_2$ in argon as the reductant.

Results and Discussion

Cu-ZrO$_2$. As reported earlier *(6,7)*, coprecipitated Cu-ZrO$_2$ catalysts were active for the selective reduction of NO to N$_2$ by propene. The activity for N$_2$ production on all the Cu-ZrO$_2$ samples showed a volcano-shaped dependence on temperature, with a broad maximum located at temperatures where the propene conversions were close to completion. This behavior was similar to those observed on many of the selective NO reduction catalysts such as Cu-ZSM-5 *(1)* and Cu/SiO$_2$-Al$_2$O$_3$ *(12)*.

The activity, and thus the temperature of maximum NO conversion, and the % NO competitiveness depended on the Cu loading (Table I). As the Cu loading was increased, both the temperature of maximum NO conversion and the % NO competitiveness decreased. This indicates that increased activity for both propene oxidation and NO reduction occurred with increased Cu loading, and that the propene oxidation activity increased faster than the NO reduction activity, resulting in lower efficiency in converting NO to N$_2$. Under comparable conditions (space velocity 10,400 h^{-1}, 642 K), a 3.2 wt.% Cu-ZSM-5 catalyst gave 72% NO conversion, 100% propene conversion, and 8.3% NO competitiveness *(6)*. Thus the performance of the most effective Cu-ZrO$_2$ sample (2.1 wt.% Cu-ZrO$_2$) was comparable to that of Cu-ZSM-5. The higher competitiveness in utilizing NO to oxidize propene on lower Cu loading samples was also observed at low conversions. This is shown in Table II where the data were obtained by extrapolation of the results in Fig. 1. These data also show that the % NO competitiveness decreased with increasing temperature on both samples. It is interesting to note that, on the 2.1 wt.% Cu-ZrO$_2$ at 601 K, the % NO competitiveness at low conversion and for high conversion are comparable.

This reaction trend for Cu-ZrO$_2$ is consistent with the model proposed earlier *(6)*, that highly dispersed Cu ions are desirable for selective reduction of NO with propene. This model assumes that CuO is not efficient for NO reduction because the Cu ions are too easily reduced and the lattice oxygen are very reactive

Table I: NO Conversion and % Competitiveness as a Function of Cu Loading for Cu-ZrO$_2$

Wt. % Cu	W/F g-min/cc	SV h^{-1}	BET m^2/g	T$_{max}$[a] K	% NO[a] Conv.	% C$_3$H$_6$[a] Conv.	% NO[b] Compet.
Cu-ZrO$_2$[c]							
0.0	0.010	13300	167.6	784[d]	25[d]	34[d]	8.4[d]
2.1	0.010	13100	163.5	601	68	91	8.6
6.0	0.014	9680	92.2	561	58	92	7.7
7.6	0.0098	13600	138.3	538	64	90	8.3
8.9	0.010	13300	157.8	536	58	81	7.9
11	0.011	12600	151.5	518	54	82	7.9
15	0.0094	14400	160.5	511	52	91	6.8
24	0.0098	13800	117.5	511	46	90	5.9
33	0.010	13000	135.9	504	43	91	5.5
Cu-ZSM-5[e]							
3.2	0.0027	10400	-	642	72	100	8.3

a.   At T$_{max}$, temperature of maximum NO conversion.
b.   % NO competitiveness at T$_{max}$.
c.   Feed: 0.1% NO, 0.1% propene, 1% O$_2$, balance He.
d.   Not at T$_{max}$. T$_{max}$ > 784 K.
e.   Feed: 0.1% NO, 0.1% propene, 2.5% O$_2$, balance He. Sample provided by General Motors Corporation.

such that the reaction of adsorbed propene and hydrocarbon intermediates with lattice oxygen competes effectively with adsorbed oxygen and adsorbed NO, thereby lowering the efficiency of the reductants to remove NO. Therefore, reducing the reactivity of lattice oxygen for hydrocarbon oxidation would increase the efficiency of NO removal. One method to achieve this is to disperse Cu ions in a nonreducible matrix, such as ZrO$_2$. According to the model, the competitiveness for NO utilization would decrease with increasing Cu loading because of increasing extent of clustering of Cu–O species which results in an increase in the amount of reactive lattice oxygen ions.

The effect of Cu loading on the apparent activation energy of NO conversion to N$_2$ and propene conversion was also examined. Fig. 1 shows the Arrhenius plots for propene and NO reaction for the 2.1 wt.% Cu-ZrO$_2$ and 33 wt.% Cu-ZrO$_2$. For the 2.1 wt.% sample, the apparent activation energy for NO conversion to N$_2$ was 10±3 kcal/mol, and that for C$_3$H$_6$ conversion was 17±3 kcal/mol. For the 33 wt.% sample, they were slightly higher, being 14±3 kcal/mol

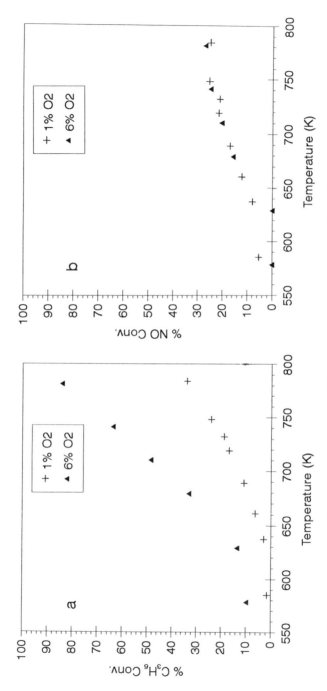

Fig. 1.   Arrhenius plots of propene and NO reaction rates for a) 2.1 wt. % and b) 33 wt. % Cu-ZrO$_2$.

Table II. Comparison of % NO Competitiveness as a Function of Temperature for 2.1 wt.% and 33 wt.% Cu-ZrO$_2$. (1000 ppm NO, 1000 ppm C$_3$H$_6$, 2% O$_2$)

Catalyst wt.% Cu	Temp. K	% NO Compet.[a]
2.1	504	14
33	504	3.9
2.1	553	7.5
33	553	2.3
2.1	601	4.5
	601	5.6[b]
33	601	1.5

a.   Determined from the Arrhenius plots of data obtained at low NO and C$_3$H$_6$ conversions.

b.   Determined from high conversion data at the temperature of maximum NO conversion.

and 20±3 kcal/mol, respectively. It is interesting to note that the values for propene conversion were higher than for NO conversion on the catalysts. In contrast, Bennett et al. *(13)* reported that for a 5 wt.% Cu-ZSM-5 catalyst, the apparent activation energies for NO and propene reactions were 25±1 kcal/mol and 11.2±0.7 kcal/mol, respectively. That is, the value for NO conversion was much higher than that for propene conversion.

The reason that the activation energy for NO conversion appeared to be lower than that for propene on the Cu-ZrO$_2$ systems is because the efficiency of using the activated hydrocarbon decreased with elevation in temperature (Table II). Since reaction of lattice oxygen with surface hydrocarbon intermediates involves breaking strong bonds in the lattice, it would have higher activation energies (i.e. stronger temperature dependence) than reactions involving adsorbed NO or adsorbed oxygen. The fact that the activation energy for NO conversion is higher than that for C$_3$H$_6$ conversion on Cu-ZSM-5 implies a higher NO competitiveness at higher temperatures.

EPR spectroscopic data of samples in Table I provided evidence that the dispersion of Cu ions decreased with increasing Cu content. Figure 2 shows the EPR signals of some of the samples. The spectrum of the 2.1 wt.% sample was quite symmetric and relatively narrow. As the Cu loading increased, the spectrum became broader and it was increasingly evident that the spectrum did not return to the baseline, suggesting the presence of a very broad feature. The broadness was attributed to an increase in the extent of Cu-Cu nuclear spin-spin interactions with increasing Cu loading. For CuO, the nuclear spin-spin interaction was so strong that the EPR signal was too broad to be observable. The EPR data of

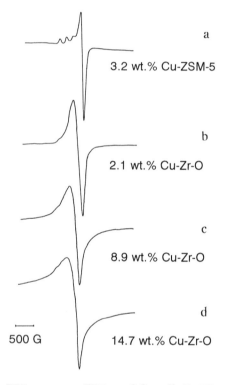

a

3.2 wt.% Cu-ZSM-5

b

2.1 wt.% Cu-Zr-O

c

8.9 wt.% Cu-Zr-O

d

500 G   14.7 wt.% Cu-Zr-O

Fig. 2.   *EPR spectra at 77K.   a: 3.2 wt.% Cu-ZSM-5; b: 2.1 wt.%*
Cu-ZrO$_2$;   c: 8.9 wt.% Cu-ZrO$_2$;   d: 14.7 wt.% Cu-ZrO$_2$.

these Cu-ZrO$_2$ samples, when coupled with the XPS data *(6)* that showed no evidence of preferential segregation of Cu on the surface of these samples, suggested that the fraction of surface Cu ions in a highly dispersed state decreased with increasing Cu content. However, even for the 2.1 wt.% Cu-ZrO$_2$ sample, the well defined hyperfine structure observed for Cu^{2+} ions in Cu-ZSM-5 *(14)* was not observed. Thus, although well dispersed, the Cu^{2+} ions in this sample were sufficiently close to have some nuclear spin-spin interaction.

The TPR data also supported this picture. Fig. 3 shows that the TPR profile for the 7.4 wt.% Cu-ZrO$_2$ sample contained only one reduction peak, which was assigned to the reduction of well dispersed Cu ions. For the higher Cu loading samples, a higher temperature reduction peak appeared which increased in size with increasing Cu loading. The spectrum of the 33 wt.% Cu-ZrO$_2$ sample is shown in Fig. 3. The higher temperature peak could be assigned to reduction of clusters of Cu–O units or small crystallites of CuO. These data were similar to those reported by Amenomiya et al. *(10)*. The amount of hydrogen consumed in the TPR for both samples corresponded to complete reduction of Cu^{2+} to its zero valent state.

Cu-Ga$_2$O$_3$. The Cu ions in a 1.9 wt.% Cu-Ga$_2$O$_3$ sample, with a surface area of 48.3 m^2/g, were also well dispersed as indicated by the Cu^{2+} EPR signal of the sample. Similar to the 2.1 wt.% Cu-ZrO$_2$ sample, the g$_{||}$ hyperfine splitting was not well resolved, suggesting that most of the Cu ions were not isolated. The catalyst was effective in the catalytic reduction of NO to N$_2$ with propene under the lean condition. The data in Table III compare the catalytic behavior of Cu-Ga$_2$O$_3$ and Cu-ZrO$_2$. Under comparable conditions, Cu-Ga$_2$O$_3$ was more effective than Cu-ZrO$_2$. In particular, the decline in NO conversion for Cu-Ga$_2$O$_3$ was less rapid with increasing O$_2$ partial pressure than for Cu-ZrO$_2$.

ZrO$_2$ and Ga$_2$O$_3$. One possible explanation for the observed differences in the catalytic behavior of Cu-ZrO$_2$ and Cu-Ga$_2$O$_3$ was that the matrices participated in the overall steady state reaction in varying degrees and that the % NO competitiveness was different for reactions that took place over the matrices and the Cu active sites. Fig. 4a and 4b are temperature profiles of C$_3$H$_6$ and NO conversions, at 1% and 6% O$_2$ in the feed, over ZrO$_2$. Propene conversion was a strong function, whereas NO conversion was apparently independent, of O$_2$ concentration in the feed. The latter is probably a coincidental consequence of the decrease in the % NO competitiveness being offset by the increase in hydrocarbon conversion with higher O$_2$ partial pressure. The NO conversion was only noticeable above 625 K and was practically zero at temperatures where Cu-ZrO$_2$ catalysts usually operate (Table I).

Table IV shows the catalytic data for Ga$_2$O$_3$ at different O$_2$ concentrations in the feed at 709 K. After reaction in a feed containing 1% O$_2$, Ga$_2$O$_3$ turned from white to a light tan with black specks interspersed in between. The 1.9 wt.% Cu-Ga$_2$O$_3$ catalyst was light green after reaction, suggesting that Cu helped in preventing carbon deposition. Unlike ZrO$_2$, the NO conversion for Ga$_2$O$_3$ improved with higher O$_2$ concentration and was 100% (with ≥90% selectivity to

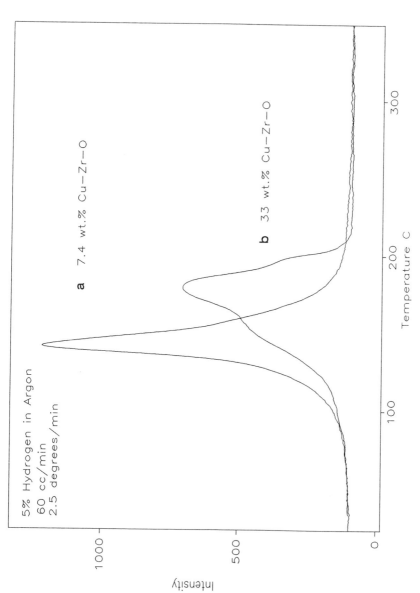

Fig. 3.   TPR profiles of Cu-ZrO₂ samples.   a: 7.4 wt.% Cu-ZrO₂; b: 33 wt.% Cu-ZrO₂.   Heating rate = 2.5 K/min.

Table III. Reduction of NO with Propene over 2.1 wt.% $Cu-ZrO_2$ and 1.9 wt.% $Cu-Ga_2O_3$. (1000 ppm N0, 1000 ppm C_3H_6).

Catalyst[d]	$\%O_2$	W/F^a (SV)[b]	T_{max} K^c	% NO Conv.	% NO Compet.
$Cu-ZrO_2$	1.0	0.01(13,100)	601	68	8
$Cu-ZrO_2$	3.1	0.01(13,100)	601	38	5
$Cu-ZrO_2$	6.2	0.01(13,100)	601	20	2
$Cu-Ga_2O_3$	1.0	0.01(6,100)	621	77	10
$Cu-Ga_2O_3$	4.1	0.01(6,100)	621	43	6
$Cu-Ga_2O_3$	6.7	0.01(6,100)	621	29	3

a.　g-min/cc.
b.　Space velocity, h^{-1}.
c.　Temperature of maximum NO conversion determined when O_2 was 1%.
d.　BET area was 163.5 m^2/g for $Cu-ZrO_2$ and 48.3 m^2/g for $Cu-Ga_2O_3$.

Table IV. Selective Reduction of NO over $Ga_2O_3^a$. (1000 ppm NO, 1000 ppm C_3H_6, 709 K^b, W/F=0.01g-min/cc)

$\%O_2$	% NO Conversion	% NO Compet.	$\% C_3H_6$ Conv.
1.0	55	10	63
4.0	90	12	84
6.7	91	11	92

a.　Surface area = 38 m^2/g.
b.　This was the temperature of maximum NO conversion determined for a 6.7% O_2 feed.

N_2, rest to N_2O) at O_2 ≥4%. Comparing the data in Table IV with those for Cu-Ga_2O_3 in Table III, it can be seen that the Ga_2O_3 was less active than $Cu-Ga_2O_3$. Its temperature of maximum NO conversion was 88 K higher. However, the data in Table V, which compare the NO reduction activity, at 1% O_2 in the feed, of Ga_2O_3 and 2.4 wt.% $Cu-Ga_2O_3$ at 648 K (near the temperature of maximum NO conversion for $Cu-Ga_2O_3$) show that even at this low temperature, the rate of N_2 production over the matrix was significant compared with the Cu supported catalyst. In addition, the data in Tables III and IV show that Ga_2O_3, relative to $Cu-Ga_2O_3$, had much higher % NO competitiveness at higher O_2 partial pressure. Thus, reaction on the Ga_2O_3 matrix plays an important role in the overall N_2 production on the $Cu-Ga_2O_3$ sample, particularly at high O_2 partial pressures.

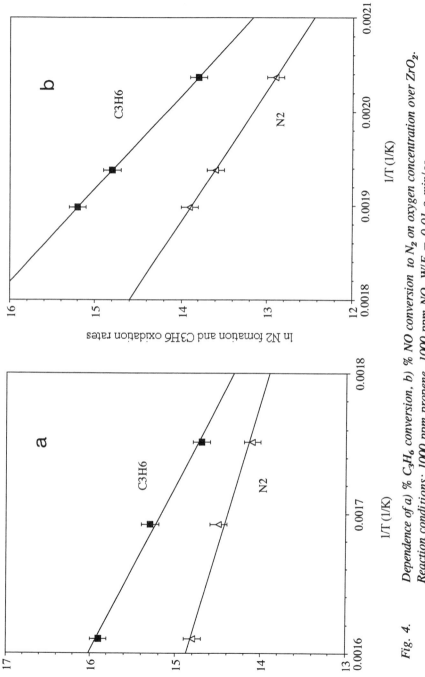

Fig. 4. Dependence of a) % C_3H_6 conversion, b) % NO conversion to N_2 on oxygen concentration over ZrO_2. Reaction conditions: 1000 ppm propene, 1000 ppm NO, W/F = 0.01 g-min/cc.

NO vs. NO_2. The above discussion over the degree of participation of the matrix in the steady state reaction did not take into account NO_2 formation over the catalysts and the different reactivity of NO_2, as compared with NO, towards the reductant hydrocarbon. It has been observed that the oxidation of NO to NO_2 is rapid over some catalysts in conditions similar to those used in NO reduction in excess oxygen *(15,16)*. The effectiveness of Cu in catalyzing NO_2 formation cannot be properly assessed from the data of Hamada and coworkers *(15)* as the reaction conditions they used were such that close to equilibrium distribution of NO_2 and NO were achieved over both Cu-ZSM-5 and H-ZSM-5. However, Arai et al. *(17)* and Karlsson et al. *(18)* demonstrated that Cu in 13X zeolite was responsible for catalyzing NO_2 formation from NO and O_2. Shimokawabe et al. *(19)* have shown that at 773 K, the areal rate of NO_2 decomposition was $CuO >> ZrO_2 > Al_2O_3 >>> SiO_2$. Thus, if one of the consequences of adding Cu to an inert oxide matrix was to increase the NO_2/NO ratio in the gas phase, then the contribution of the matrix to the overall catalytic reaction may be altered by the increased level of NO_2.

It has been found that on Cu-ZSM-5 using isobutane as the reductant, the reduction of NO_2 was slightly faster than the reduction of NO *(20)*. Iwamoto et al. *(21)* found that for the selective reduction of NO by ethene over Cu-ZSM-5 catalysts, NO_2 was more effective than NO between 350-500 K. Thus, it is possible that NO reduction on Cu-ZrO_2 and Cu-Ga_2O_3 proceeded via first the formation of NO_2.

In order to evaluate this possibility, the rates of reduction of NO and NO_2 with propene were compared over Ga_2O_3 and ZrO_2. For the purpose of ensuring identical reaction conditions, experiments were conducted on the same apparatus

Table V. Comparison of Rates of Reduction of NO and NO_2. (Feed = 1000 ppm NO_x, 1000 ppm propene, 1% O_2).

Catalyst	Reactant	Temp. K	NO_x Conv.%	% NO Compet.	Propene Conv.%
ZrO_2	NO	598	0	-	0
	NO_2[a]	598	24	28	10
2.1 wt.%- Cu-ZrO_2	NO	601	66	8	97
Ga_2O_3	NO	648	21	14	17
	NO_2[a]	648	58	22	30
1.9 wt.%- Cu-Ga_2O_3	NO	632	77	9	92

a.   Feed contained an equilibrium mixture of 680 ppm NO_2, 320 ppm NO.

using the same feed of NO, O_2, and propene, except that in the experiments with NO_2, the NO and O_2 stream was first passed through a reactor containing 1 g Co/Al_2O_3 catalyst at 573 K to achieve the equilibrium composition of N_2O/NO at that temperature. The conversion of NO to NO_2 over the Co/Al_2O_3 catalyst was independently confirmed by monitoring the gas mixture with Fourier transform infrared spectroscopy. The resulting mixture was then mixed with propene and passed over the ZrO_2 or Ga_2O_3 catalyst. Thus, the feed to these matrix oxides consisted of 1000 ppm propene, 1% O_2, and about 680 ppm NO_2 and 320 ppm NO in He.

On both Ga_2O_3 and ZrO_2, the rate of formation of N_2 from NO_2 was much higher than from NO (Table V). In fact, these rates suggested that the contribution of the matrix to the overall N_2 production would be substantial if the Cu in the mixed oxide catalysts had catalyzed the oxidation of NO to the equilibrium ratio of NO_2/NO. However, our IR data suggested that $Cu-Ga_2O_3$ was not a very effective catalyst for the oxidation of NO to NO_2. Even in the presence of 4% O_2, the overall NO conversion to NO_2 was only about 40% at 573 K. Thus, the experiments in which efficient NO_2 formation was ensured by pre-oxidation of NO on Co/Al_2O_3 before entering the reactor placed an estimate of the upper limit of the degree of participation of the matrix in the production of N_2 over a Cu-containing catalyst.

Summary and Conclusion

It was demonstrated in this study that Cu ions highly dispersed in matrices that are not active in propene combustion, such as ZrO_2 and Ga_2O_3, are active catalysts for propene reduction of NO to N_2 in the lean condition. For the $Cu-ZrO_2$ catalysts, it was found that increasing the Cu content resulted in catalysts that were more active in both propene combustion and NO reduction. But the increase in propene combustion activity was faster than NO reduction, thus the resulting efficiency in NO removal decreased. EPR spectra and TPR profiles showed that catalysts with higher Cu contents contained larger fraction of Cu ion in clusters.

Under similar conditions, $Cu-Ga_2O_3$ was more efficient in NO reduction than $Cu-ZrO_2$. It was also less sensitive to suppression by high oxygen partial pressure. The differences in the two catalysts may be due to the degree of participation of the matrices in N_2 production relative to Cu. In the absence of efficient oxidation of NO to NO_2, ZrO_2 was considered too inactive to contribute to the overall N_2 production at the temperatures where $Cu-ZrO_2$ was active. However, the Ga_2O_3 matrix contributed to the overall N_2 production at temperatures where $Cu-Ga_2O_3$ operated. NO_2 was a much more effective oxidizer of the hydrocarbon than NO over both Ga_2O_3 and ZrO_2. If NO in the feed could be efficiently oxidized to NO_2, then even ZrO_2 would contribute to the overall NO_x conversion on $Cu-ZrO_2$.

Results of this study suggest that the possible role of transition metal in these mixed oxide catalysts are to: 1) prevent coking, 2) generate NO_2, and 3) form N_2. The role of a desirable matrix oxide is to disperse the metal function as well as participate in N$_2$ production. Synergism results when NO$_2$ produced on the transition metal enhances the contribution of the matrix oxide in N$_2$ production.

Acknowledgments

This work was supported by Engelhard Industries, General Motors Corporation and NSF grant CTS-9308465. We thank Professor Can Li for testing the NO$_2$ formation ability of some catalysts with IR.

Literature Cited

1. Held, W.; Konig, A.; Richter, T.; Puppe, L. *SAE Paper*, **1990,** no. 900 496.
2. Iwamoto, M.; Yahiro, H.; Shundo, S.; Yu-u, Y.; Mizuno, N. *Shokubai*, **1990,** *33*, 430.
3. Iwamoto, M.; Hamada, H. *Catal. Today*, **1991,** *10*, 57.
4. Treux, T.; Searles, R.; Sun, D. *Platinum Metals Rev.* **1992,** *36*, 2.
5. Grinsted, R.; Jen, H.; Montreuil, C.; Rokosz, M.; Shelef, M. *Zeolites*, **1993,** *13*, 602.
6. Bethke, K.; Alt, D.; Kung, M. *Catal. Lett.* **1994,** *25*, 37.
7. Kung, M.; Bethke, K.; Kung, H. *Preprint ACS Div. Petrol. Chem.* **1994,** *39*, 154.
8. Ansell, G.; Diwell, A.; Golunski, S.; Hayes, J.; Rajaram, R.; Treux, T.; Walker, A. *presented in 205th ACS National Meeting, Denver, CO, 29 March - 1 April*, **1993**.
9. Montreuil, C.; Gandhi, H.; Chatta, M. *U.S. Patent*, **1992,** 5,155,077.
10. Amenomiya, Y.; Ali Emesh, I.; Oliver, K.; Pleizer, G. in *Proc. 9th Intern. Cong. Catal.* Editors Phillips, M.; Ternan, M. Chemical Institute of Canada, Ottawa, Canada, **1988,** pp. 634.
11. Cho, B. *J. Catal.* **1993,** *142*, 418.
12. Hosose, H.; Yahiro, H.; Mizuno, N.; Iwamoto, M. *Chem. Lett.* **1991,** 1859.
13. Bennett, C.; Bennett, P.; Golunski, S.; Hayes, J.; Walker, A. *Appl. Catal. A*, **1992,** *86*, L1.
14. Kucherov, A.; Slinkin, A.; Kondratiev, D.; Bondarenko, T.; Rubinshtein, A.; Minachev, Kh. *Kinet. Katal.* **1985,** *26*, 353.
15. Sasaki, M.; Hamada, H.; Kintaichi, Y.; Ito, T. *Catal. Lett.* **1992,** *15*, 297.
16. Hamada, H.; Kintaichi, Y.; Ito, T.; Sasaki, M. *Appl. Catal.* **1991,** *70*, L15.
17. Arai, H.; Tominaga, H.; Tsuchiya, J. in *Proc. 6th Intern. Cong. Catal.* Letchworth, England, **1977,** pp. 997.
18. Karisson, H.; Rosenberg, H. *Ind. Eng. Chem. Process Des. Dev.* **1984,** *23*, 804.
19. Shimokawabe, M.; Ohi, A.; Takezawa, N. *Appl. Catal. A*, **1992,** *85*, 129.
20. Petunchi, J.; Hall, W.K. *Appl. Catal. B*, **1993,** *2*, L17.
21. Iwamoto, M.; Mizuno, N. *Preprints, Part D Proc. Institute of the Mechanical Engineers, J. Auto. Engin.* **1992**.

RECEIVED November 7, 1994

Chapter 9

IR Study of Catalytic Reduction of NO_x by Propene in the Presence of O_2 over CeZSM-5 Zeolite

H. Yasuda[1,2], T. Miyamoto[1], and M. Misono[1]

[1]Department of Synthetic Chemistry, Faculty of Engineering, University of Tokyo, Bunkyo-ku, Tokyo 113, Japan

The adsorbed species formed on Ce-ZSM-5 during the catalytic reduction of nitrogen monoxide (NO) by propene (C_3H_6) in the presence of oxygen using a closed circulation system and their reactivities have been investigated by means of infrared spectroscopy coupled with the quantitative analysis of the gas phase. Organic nitro- (1558 cm^{-1}), nitrito- (1658 cm^{-1}) compounds, and isocyanate (2266 and 2241 cm^{-1}) species were formed in addition to carbonyl, nitrate, etc. on the surface during the NO_2-C_3H_6-O_2 reaction at 373 K, and the band due to the nitro-compounds rapidly decreased correspondingly to the formation of N_2 in the gas phase when the temperature was raised to 423 and 473 K in the presence of NO_2. The weak band due to isocyanate species (2241 cm^{-1}) also decreased gradually, but the change was in general much smaller. Similar behavior was observed for the reaction between adsorbed CH_3NO_2 and NO_2. Based on these results, it was presumed that organic nitro-compounds formed rapidly from NO_2 and C_3H_6 were possible intermediates and that N_2 was mainly produced by the reactions between the nitro-compounds and NO_2 (or NO_2 and O_2) in the NO-C_3H_6-O_2 reaction over Ce-ZSM-5.

The catalytic removal of nitrogen oxides (NO_x) is a technology urgently required for the protection of our atmospheric environment. The catalytic reduction of nitrogen

[2]Current address: Chemical Technology Division, Institute of Research and Innovation, Takada 1201, Kashiwa-shi, Chiba 227, Japan

monoxide (NO) by hydrocarbons in the presence of excess oxygen has recently attracted much attention for the depollution of exhaust gases from diesel or lean-burn gasoline engines. Various catalysts such as zeolites (*1-12*), metal oxides (*13,14*), metallosilicates (*15,16*), and noble metals (*2,17,18*) have been reported to be active for this reaction. We previously found that rare earth ion-exchanged zeolites, especially Ce-ZSM-5, showed a high catalytic activity (*19*), and that the increase of cerium content or the addition of alkali earth metals further enhanced the activity (*20*). However, the mechanism of this reaction, particularly the nature of the reaction intermediates formed on the surface, has not yet been elucidated.

From the results of infrared spectroscopic measurements, it has been suggested that isocyanate species (NCO) was an intermediate of the reaction on Cu-Cs/Al$_2$O$_3$ (*21,22*) and Cu-ZSM-5 (*23*). We recently reported the formation of organic nitro- and nitrito-compounds from NO$_2$ and C$_3$H$_6$ over Pt/SiO$_2$ (*24*) and Ce-ZSM-5 (*25,26*). In the present study, we investigated the adsorbed species formed from NO$_2$-C$_3$H$_6$-O$_2$ and their reactivities on Ce-ZSM-5 catalyst by means of infrared spectroscopy coupled with the quantitative analysis of the gas phase.

Experimental

Ce ion-exchanged ZSM-5 used in this study was prepared in the following way by the repeated ion-exchange at 298 K of parent ZSM-5 zeolite (SiO$_2$/Al$_2$O$_3$ = 23.3) supplied by Tosoh Corporation. About 20 g of parent ZSM-5 was first ion-exchanged in an aqueous sodium nitrate solution, washed with water, and dried to obtain Na-ZSM-5, which was then ion-exchanged in 1.2 dm^3 of aqueous cerium acetate solution (0.1 mol·dm^{-3}) for two days. The zeolite was then filtered, washed with water, and dried at 393 K overnight. Ce-ZSM-5 thus obtained was ion-exchanged again in a new cerium acetate solution in order to increase the cerium content in the zeolite. The ion-exchange level was estimated by the amount of the eluted sodium ion, which was measured with atomic absorption spectroscopy by assuming that cerium was present as Ce^{3+}. The exchange level of Ce-ZSM-5 used in this study was 18.5 % (1.3 wt% as CeO$_2$). The powder X-ray diffraction pattern of the ion-exchanged Ce-ZSM-5 was the same as that of Na-ZSM-5, no diffraction peaks due to CeO$_2$ being observed.

Infrared spectra were collected by using an FTIR-8500 spectrometer (Shimadzu Co., Ltd.) and an IR cell (with CaF$_2$ windows) connected to a closed circulation system (ca. 620 cm^3). The catalyst was pressed into a wafer (ca. 40 mg), and supported by a glass-made holder which could be moved up and down in the cell to bring the wafer in and out of the optical path. A small basket which contained about 60 mg of catalyst was placed near the catalyst wafer but out of optical path in order to increase the rate of reaction. Thus, it was possible to measure the IR spectra and the composition of the gas phase simultaneously. The catalyst was evacuated at 773 K for 30 min, exposed to oxygen (P$_{O2}$ = 100 Torr, 1 Torr = 0.133 kPa) for 30 min, evacuated again for 30 min, cooled to a desired temperature, and then exposed to reactant gases. IR spectra from the adsorbed species were obtained by subtracting the spectrum of the pretreated wafer and the gas phase at each temperature from the spectrum obtained after adsorption or reaction. Products in the gas phase were analyzed by a gas chromatograph which was directly connected to the circulation system.

Results and Discussion

Adsorbed Species Formed from NO$_2$-C$_3$H$_6$-O$_2$ at 373 K.

Figure 1 (A) and (B) shows the IR spectra of the adsorbed species formed after the introduction of the gas

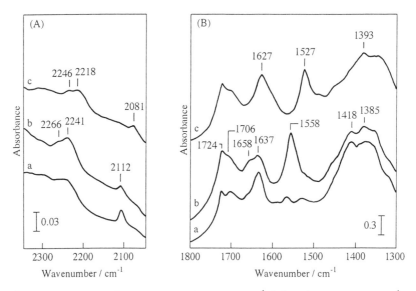

Figure 1. IR spectra in the range 2350 - 2050 cm^{-1} (A) and 1800 - 1300 cm^{-1} (B) of the adsorbed species formed on Ce-ZSM-5 at 373 K after exposure to the ^{14}NO (10 Torr)-C$_3$H$_6$ (10 Torr)-O$_2$ (50 Torr) mixed gas for 1 min (a), 60 min (b), and ^{15}NO-C$_3$H$_6$-O$_2$ for 59 min (c) (note that the actual composition of reactant was close to NO$_2$ (10 Torr)-C$_3$H$_6$ (10 Torr)-O$_2$ (45 Torr), see text).

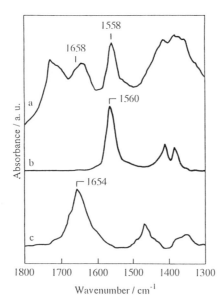

Figure 2. IR spectra of the adsorbed species formed on Ce-ZSM-5 during the NO$_2$-C$_3$H$_6$-O$_2$ reaction at 373 K (a), the adsorbed CH$_3$NO$_2$ on Ce-ZSM-5 at 423 K (b), and the adsorbed C$_4$H$_9$ONO on SiO$_2$ at 298 K (c).

mixture of NO (10 Torr), C_3H_6 (10 Torr), and O_2 (50 Torr) on Ce-ZSM-5. Spectra a and b were obtained at 373 K after 1 min and 1 h, respectively. Several bands appeared in the range 2400 - 1300 cm^{-1}, while no significant bands were observed in the range 2100 - 1800 cm^{-1}. As shown in spectrum c, the bands at 2266, 2241, 2112, 1658, 1558, 1418, and 1385 cm^{-1} shifted to lower wavenumbers when ^{15}NO was used in place of ^{14}NO. Upon the introduction of the NO-C_3H_6-O_2 gas mixture, N_2, N_2O, CO_2 and CO were formed at a significant rate in the gas phase in the initial stage, but the rates of formation slowed down with time. In addition to these products, a small amount of HCN, which was observed in flow reactor experiments (*25,27*), was also produced in the gas phase. The band at 1558 cm^{-1}, which is the possible reaction intermediate of the N_2 formation as discussed below, developed with time.

In this experiment, NO in the gas phase was mostly oxidized to NO_2 by the coexisting oxygen, before the adsorption was initiated, as the NO-C_3H_6-O_2 gas mixture showed red-brown color. It was also confirmed by the IR spectra (NO_2(g): 1618 cm^{-1}) of the gas phase. It was previously concluded (*25*), on the basis of the comparison of the rates of NO-C_3H_6-O_2, NO_2-C_3H_6(-O_2), and NO-O_2 reactions in flow reactor experiments, that the NO-C_3H_6-O_2 reaction on Ce-ZSM-5 proceeded via the reaction between NO_2 and C_3H_6. Hence, the NO_2-C_3H_6-O_2 reaction studied in a closed system in this study would reasonably reflect the NO-C_3H_6-O_2 reaction in flow experiments.

The adsorbed species formed from C_2H_4 [NO_2 (10 Torr)-C_2H_4 (9 Torr)-O_2 (45 Torr)] and C_3H_8 [NO_2 (10 Torr)-C_3H_8 (10 Torr)-O_2 (45 Torr)] were also investigated. The IR spectra of the adsorbed species for C_2H_4 were similar to those for C_3H_6, except that the bands at 2266 and 2241 cm^{-1} were much smaller for C_2H_4. On the other hand, the IR spectra were quite different in the case of C_3H_8; bands observed for the adsorption of NO_2 alone were dominant. The amounts of N_2 and N_2O formed were in the order of $C_3H_6 > C_2H_4 \gg C_3H_8$.

Assignments of the IR Bands. The bands of spectra a and b in Figure 1 are assigned as follows. As shown in spectrum b in Figure 2, CH_3NO_2 adsorbed on Ce-ZSM-5 exhibited a band at 1560 cm^{-1}, which was assigned to ν_{as} (NO_2) (*24,28*). A band assignable to ν (NO) appeared at 1654 cm^{-1} for adsorbed C_4H_9ONO on SiO_2 (spectrum c in Figure 2). Therefore the bands at 1558 and 1658 cm^{-1} of spectra a and b in Figure 1 are attributed respectively to ν_{as} (NO_2) and ν (NO) of organic nitro- and nitrito-compound(s) formed by the reaction between NO_2 and C_3H_6. Although the detailed structure of these nitro- and nitrito-compounds are not clear yet, the formation of those compounds is known in the reactions between alkenes and NO_2 (*29*). These bands did not appear when only NO_2 was adsorbed. Iwamoto, et al., (*23*) have reported two isocyanate (NCO) bands at 2189 and 2251 cm^{-1} for Cu-ZSM-5 and assigned them to NCO on Cu and on the zeolite framework, respectively. So, the small bands at 2266 and 2241 cm^{-1} of spectra a and b in Figure 1 are probably due to two NCO species (note that the scale of Figure 1 (A) is one tenth that in Figure 1 (B)). The band at 2112 cm^{-1} may be due to adsorbed cyanide (*30*) or NO_2 (*31*). The broad doublet at 1418 and 1385 cm^{-1} is assigned to nitrate (*30*). The bands at 1724 and 1706 cm^{-1} are tentatively assigned to carbonyl species formed under the reaction conditions, since the position of these bands did not shift when ^{15}NO was used, and carbonyl species have a characteristic band at around 1700 cm^{-1}. The band at 1637 cm^{-1} is due to adsorbed water.

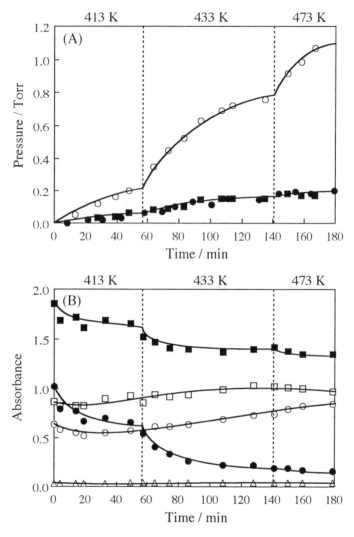

Figure 3. Changes of the amounts of N$_2$ (●), N$_2$O (■) and CO$_2$ (○) formed
(A) and the absorbances of the bands at 2241 (△), 1724 (□), 1658 (○), 1558
(●), and 1418 (■) cm^{-1} (B) during the elevation of the temperature in a closed
system after the NO$_2$ (10 Torr)-C$_3$H$_6$ (10 Torr)-O$_2$ (45 Torr) reaction over Ce-
ZSM-5 at 373 K and subsequent evacuation.

Thermal Desorption of the Adsorbed Species. The spectrum b in Figure 1 little changed upon evacuation at 373 K, except that the band at 2112 cm^{-1} disappeared. When the temperature was raised after evacuation to 473 K in a closed system, N$_2$, N$_2$O, and CO$_2$ evolved in the gas phase, as shown in Figure 3 (A). Concomitantly, the band at 1558 cm^{-1} and the doublet at around 1400 cm^{-1} decreased, while the bands at 2266, 2241, 1724, 1706, and 1658 cm^{-1} little changed (Figure 3 (B)). In this experiment, it took about 3 min to increase the temperature by 50 degrees, and the vertical dotted lines in Figure 3 (A) and (B) show the time when the desired temperature was reached.

As shown in Figure 3 (A) and (B), good correspondences exist between the decrease of the band at 1558 cm^{-1} and the increase of N$_2$, N$_2$O, and CO$_2$. By contrast, the correlations were not found for the bands at 1658 cm^{-1} (nitrito-compound), 2266 and 2241 cm^{-1} (NCO), or 1724 cm^{-1} (carbonyl). The doublet at around 1400 cm^{-1} (nitrate) also decreased. However, the relative change of the band intensity was smaller than that of the band at 1558 cm^{-1} and moreover it was confirmed by a separate experiment that N$_2$ was not produced by thermal desorption of nitrate species which was formed only from NO$_2$. Therefore, it is unlikely that N$_2$ observed in the above experiment was produced from nitrate. Thus, N$_2$, N$_2$O, and CO$_2$ were most probably produced by the decomposition of the nitro-compounds formed during the NO$_2$-C$_3$H$_6$-O$_2$ reaction.

Reactivities of the Adsorbed Species with NO$_2$, O$_2$, or NO. After the NO$_2$-C$_3$H$_6$-O$_2$ reaction at 373 K for 1 h and subsequent evacuation as in the above experiment, the temperature was raised in the presence of NO$_2$ (10 Torr), O$_2$ (50 Torr), or NO (10 Torr) to 473 K, and then kept for 1 h at 473 K. The amounts of N$_2$, N$_2$O, CO$_2$, and CO produced in the gas phase during this experiment are summarized in Table I, together with the atomic ratios of nitrogen to carbon (N/C) in the gaseous products. The amounts of products formed by the experiment described in the preceding section are also included in the table for comparison. As shown in Table I, the reactivity of

Table I. Products Formed in the Gas Phase during the Reaction of the Adsorbed Species Formed from NO$_2$-C$_3$H$_6$-O$_2$ with NO$_2$, O$_2$ or NO

Reactions	Products / Torr[a]				N/C[b]
	N$_2$	N$_2$O	CO$_2$	CO	
NO$_2$ (10 Torr)[c]	1.6	0.5	3.2	1.1	1.0
O$_2$ (50 Torr)	0.7	0.1	1.3	0.3	1.0
NO (10 Torr)	0.4	0.4	1.0	0.5	1.1
_[d]	0.2	0.1	1.0	n.d.[e]	0.6

[a]Amounts of N$_2$, N$_2$O, CO$_2$, and CO formed when the temperature was increased in the presence of NO$_2$, O$_2$, NO or in vacuum and then kept at 473 K for 1 h in the same atmosphere after the NO$_2$-C$_3$H$_6$-O$_2$ reaction at 373 K over Ce-ZSM-5.

[b]Atomic ratios of nitrogen to carbon (N/C) in the products.

[c]Figures in parentheses are the partial pressure.

[d]Without the addition of NO$_2$, O$_2$, or NO.

[e]n.d. means that CO was not detected by gas chromatography.

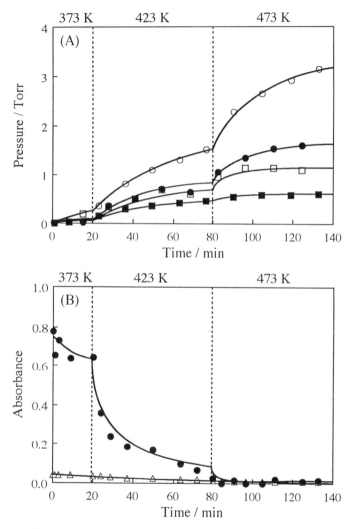

Figure 4. Changes of the amounts of N$_2$ (●), N$_2$O (■), CO$_2$ (O), and CO (□) formed (A) and the absorbances of the bands at 2241 (△) and 1558 (●) cm^{-1} (B) during the elevation of the temperature in NO$_2$ (10 Torr) after the NO$_2$ (10 Torr)-C$_3$H$_6$ (10 Torr)-O$_2$ (45 Torr) reaction over Ce-ZSM-5 at 373 K and subsequent evacuation.

the adsorbed species was greatest in the presence of NO_2. It is noted that N/C ratios were approximately unity except for the experiment without the addition of NO_2, etc. The changes of the products in the gas phase and the bands at 1558 and 2241 cm^{-1} are shown for the case of NO_2 in Figure 4 (A) and (B), respectively. The vertical dotted lines in Figure 4 mean the same as in Figure 3. Upon the introduction of NO_2 at 373 K, there were little changes in the IR spectra which were very similar to the spectrum b in Figure 1. When the temperature was raised in NO_2, the band at 1558 cm^{-1} decreased remarkably at 423 K and disappeared completely at 473 K. The band at 2241cm^{-1} also decreased gradually, but the change was small. On the other hand, the bands at 2266 and 1658 cm^{-1} little varied. As shown in Figure 4 (A), N_2, N_2O, CO_2, and CO were produced in the gas phase during this period. Figure 5 shows that linear correlations approximately exist between the rates of N_2 and CO_2 formation and the absorbance of the IR band due to nitro-compounds (1558 cm^{-1}) at 423 K. The band at 1558 cm^{-1} also decreased in O_2 or NO at elevated temperatures, but the decrease of the band intensity and the formation of gaseous products were much slower in O_2 or NO than in NO_2. The band at 2241 cm^{-1} remained almost unchanged in O_2 or NO.

These results indicate that the reaction between nitro-compounds and NO_2 is the main route of the N_2 formation for the NO_2-C_3H_6-O_2 reaction on Ce-ZSM-5 at 373 - 473 K. As described above, N_2 and N_2O were formed upon the introduction of NO_2-C_3H_6-O_2 onto Ce-ZSM-5 at 373 - 423 K and the rate of their formation slowed down with time. It is noteworthy that the rate of N_2 and N_2O formation for the NO_2-C_3H_6-O_2 reaction was of the comparable order at 423 K with that observed here at 423 K (Figure 4 (A)) and the N_2/N_2O and N/C ratios were also similar, while the direct comparison must be made cautiously, since the reaction conditions are different between the two cases.

The relative slow development of the band at 1558 cm^{-1} at 373 K (spectrum a to b in Figure 1), accompanied by the decrease of the rates of N_2 and N_2O formations, may be explained as follows. The decomposition of the nitro-compounds (1558 cm^{-1}) is rapid on fresh catalyst and hence they are hardly observable, but as the decomposition activity decreases possibly due to the accumulation of polymerization or oxidation products on the surface, the band gradually develops. The accumulation may be more pronounced at low temperatures. Rapid reaction between NO_2 and C_3H_6 has already been demonstrated for Ce-ZSM-5 *(25,26)* and SiO_2 *(24)*.

It may also be possible that a part of the nitro-compounds are transformed to N_2 via isocyanate species (and/or cyanide), since the band at 2241 cm^{-1} also slowly decreased in NO_2. On the contrary, it is unlikely that nitrito-compounds participate in the formation of N_2. Slight increase of the bands at around 1700 cm^{-1} (Figure 3 (B)) suggests the possibility of the formation of carbonyl species from the nitro-compounds as intermediates. However, the detailed mechanism subsequent to the formation of the nitro-compounds is not very clear yet.

Reactivity of Adsorbed CH3NO2. To confirm the above mechanism, the reactivity of adsorbed CH_3NO_2, which is an example of organic nitro-compounds, with NO_2, O_2 or NO was examined. First, CH_3NO_2 (2 Torr) was adsorbed on Ce-ZSM-5 at 423 K and subsequently evacuated. During this procedure the band at 1560 cm^{-1} developed rapidly (see spectrum b in Figure 2), but the amounts of products detectedin the gas phase were small. Then, NO_2 (10 Torr) was introduced at 423 K. N_2, N_2O, CO_2, and CO were formed in the gas phase, accompanied by rapid

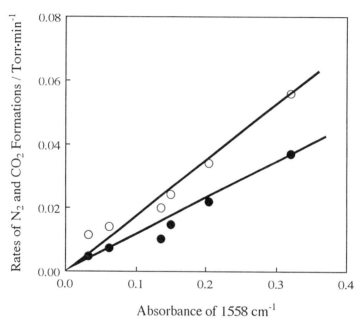

Figure 5. Correlations between the absorbance of the band at 1558 cm^{-1} and the rates of N$_2$ (●) and CO$_2$ (O) formation at 423 K in the experiment for the elevation of the temperature in NO$_2$ (10 Torr) after the NO$_2$ (10 Torr)-C$_3$H$_6$ (10 Torr)-O$_2$ (45 Torr) reaction over Ce-ZSM-5 at 373 K.

Table II. Products Formed in the Gas Phase during the Reaction of Adsorbed CH_3NO_2 with NO_2, O_2 or NO

Reactions	N_2	N_2O	CO_2	CO
		Products / Torr[a]		
NO_2 (10 Torr)[b]	0.9	0.4	0.8	0.7
O_2 (62 Torr)	0.1	0.1	0.1	n.d.[c]
NO (10 Torr)	0.1	<0.1[d]	0.1	n.d.

[a]Amounts of N_2, N_2O, CO_2, and CO formed at 423 K upon the exposure to NO_2, O_2 or NO for 1 h, after CH_3NO_2 was adsorbed on Ce-ZSM-5 at 423 K and subsequently evacuated.
[b]Figures in parentheses are the partial pressure.
[c]n.d. means that CO was not detected by gas chromatography.
[d]Amount of N_2O was less than 0.05 Torr.

decrease of the band at 1560 cm^{-1}. Those gases were also formed upon the introduction of O_2 (62 Torr) or NO (10 Torr). However, the rates and the amounts were much smaller. The amounts of the products in the gas phase are summarized in Table II. Figure 6 (A) and (B) show the changes with time of the amount of N_2 formed and the intensity of the band at 1560 cm^{-1} after the exposure to NO_2, O_2 or NO, respectively. These results are very similar to those observed for the nitro-compounds formed from NO_2-C_3H_6-O_2 and support the mechanism proposed above. When $^{15}NO_2$ was reacted with adsorbed $CH_3^{14}NO_2$, almost random isotopic distribution of ^{15}N in N_2O was observed. This result indicates the complex nature of this reaction. The reactivity of adsorbed CH_3NO_2 with the NO_2 (10 Torr) and O_2 (25 Torr) mixture was a little higher than that with NO_2 alone. So, it may be possible that oxygen in addition to NO_2 plays a role in the process of the oxidative decomposition of nitro-compounds.

Conclusion

Combination of IR study of adsorbed species with gas phase analyses showed that, in the NO_2-C_3H_6-O_2 reaction over Ce-ZSM-5 at 373 - 473 K, organic nitro-compounds were formed from NO_2 and C_3H_6, and N_2 was mainly produced by the reactions of these nitro-compounds with NO_2 (or NO_2 + O_2). These results together with the results obtained previously for flow reactor experiments indicated that these nitro-compounds are the intermediates of the reduction of NO as in the following scheme. The high reactivity of adsorbed CH_3NO_2 with NO_2 supported this scheme.

$$NO \xrightarrow{O_2} NO_2 \xrightarrow{C_3H_6} \text{organic nitro-compounds} \xrightarrow{NO_2 (+ O_2)} \to N_2, N_2O, CO_x$$

Although nitro-compounds were hardly observed by IR during the NO_2-C_3H_6-O_2 reaction above 473 K, this is possibly due to too rapid decomposition of nitro-compounds at these temperatures and it is probable that nitro-compounds are also involved in the reaction pathway at a higher temperature.

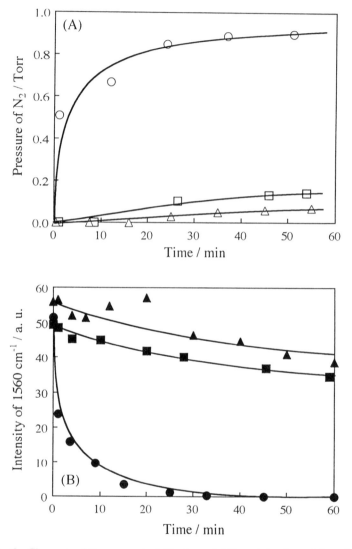

Figure 6. Changes of the amount of N$_2$ formed (A) and the intensities of the band at 1560 cm^{-1} (B) upon the exposure to NO$_2$ (10 Torr) (O,●), O$_2$ (62 Torr) (□,■) or NO (10 Torr) (△,▲), after CH$_3$NO$_2$ (2 Torr) was adsorbed on Ce-ZSM-5 at 423 K and subsequently evacuated.

Acknowledgments

The authors acknowledge Tosoh Corporation for the donation of ZSM-5 zeolite. The authors are also grateful to Mr. Yokoyama, C. and Mr. Kawamoto, M. for helpful discussion and experimental assistance. This work was supported in part by a Grant-in Aid for Scientific Research from Ministry of Education, Science and Culture of Japan.

Literature Cited

1. Iwamoto, M. *Symposium on Catalytic Technology for the Removal of Nitrogen Oxides;* Catalysis Society of Japan: Tokyo, 1990; pp 17-22.
2. Held, W.; Koenig, A.; Richter T.; Puppe, L. *SAE Paper* **1990,** 900496.
3. Hamada, H.; Kintaichi, Y.; Sasaki, M.; Ito, T.; Tabata, M. *Appl. Catal.* **1990,** *64,* L1.
4. Iwamoto, M.; Yahiro, H.; Shundo, S.; Yu-u, Y.; Mizuno, N. *Appl. Catal.* **1991,** *69,* L15.
5. Montreuil, C. N.; Shelef, M. *Appl. Catal. B* **1992,** *1,* L1.
6. Li, Y.; Armor, J. N. *Appl. Catal. B* **1992,** *1,* L31.
7. Ansell, G. P.; Diwell, A. F.; Golunski, S. E.; Hayes, J. W.; Rajaram, R. R.; Truex, T. J.; Walker, A. P. *Appl. Catal. B* **1993,** *2,* 81.
8. Burch, R; Millington, P. J. *Appl. Catal. B* **1993,** *1,* 101.
9. D'Itri, J. L.; Sachtler, W. M. H. *Appl. Catal. B* **1993,** *2,* L7.
10. Petunchi, J. O.; Hall, W. K. *Appl. Catal. B* **1993,** *2,* L17.
11. Gopalakrishnan, R.; Stafford, P. R.; Davison, J. E.; Hecker W. C.; Bartholomew, C. H. *Appl. Catal. B* **1993,** *2,* 165.
12. Kharas, K. C. C. *Appl. Catal. B* **1993,** *2,* 207.
13. Kintaichi, Y.; Hamada, H.; Tabata, M.; Sasaki, M.; Ito, T. *Catal. Lett.* **1990,** *6,* 239.
14. Teraoka, Y.; Harada, T.; Iwasaki, T.; Ikeda, T.; Kagawa, S. *Chem. Lett.* **1993,** 773.
15. Inui, T.; Iwamoto, S.; Kojo, S.; Yoshida, T; *Catal. Lett.* **1992,** *16,* 223.
16. Kikuchi, E.; Yogo, K.; Tanaka, S.; Abe, M. *Chem. Lett.* **1991,** 1063.
17. Hamada, H.; Kintaichi, Y.; Sasaki, M.; Ito, T.; Tabata, M. *Appl. Catal.* **1991,** *75,* L1.
18. Obuchi, A.; Ohi, A.; Nakamura, M.; Ogata, A.; Mizuno, K.; Ohuchi, H. *Appl. Catal. B* **1993,** *2,* 71.
19. Misono, M; Kondo, K. *Chem. Lett.* **1991,** 1001.
20. Yokoyama, C.; Misono, M. *Chem. Lett.* **1992,** 1669; *Bull. Chem. Soc. Jpn.* **1994,** *67,* 557.
21. Ukisu, Y.; Sato, S.; Muramatsu, G.; Yoshida, K. *Catal. Lett.* **1991,** *11,* 177; *Catal. Lett.* **1992,** *16,* 11.
22. Ukisu, Y.; Sato, S.; Abe, A.; Yoshida, K. *Appl. Catal. B* **1993,** *2,* 147.
23. Yahiro, H.; Yu-u, Y.; Takeda, H.; Mizuno, N.; Iwamoto, M. *Shokubai (Catalysis)* **1993,** *35,* 130.
24. Tanaka, T.; Okuhara, T.; Misono, M. *Appl. Catal. B* **1994,** *4,* L1.
25. Yokoyama, C.; Yasuda, H.; Misono, M. *Shokubai (Catalysis)* **1993,** *35,* 122.
26. Yasuda, H.; Miyamoto, T.; Yokoyama, C.; Misono, M. *Shokubai (Catalysis)* **1993,** *35,* 386.
27. Radtke, F.; Koeppel, R. A.; Baiker, A. *Appl. Catal. A* **1994,** *107,* L125.
28. Angevaare, P. A. J. M.; Grootendorst, E. J.; Zuur, A. P.; Ponec, V. In *New Developments in Selective Oxidation;* Centi, G.; Trifiro, F., Ed.; Elsevier Science Publishers B. V.: Amsterdam, 1990; pp 861-868.
29. Levy, N.; Scaife, C. W. *J. Chem. Soc.* **1946,** 1093.

30. Nakamoto, K. *Infrared Spectra of Inorganic and Coordination Compounds;* Wiley-Interscience: New York, NY, 1970.
31. Chao, C-C; Lunsford, J. H. *J. Am. Chem. Soc.* **1971,** *93,* 71.

RECEIVED November 7, 1994

Chapter 10

Mechanism of Selective Catalytic Reduction of NO by Propene on Fe Silicate in Oxygen-Rich Atmosphere

A Transient Response Study

Katsunori Yogo, Takashi Ono, Masaru Ogura, and Eiichi Kikuchi[1]

Department of Applied Chemistry, School of Science and Engineering, Waseda University, 3-4-1 Okubo, Shinjuku-ku, Tokyo 169 Japan

The reaction mechanism of NO reduction by C_3H_6 on Fe-silicate catalyst was investigated by means of a transient response method. From the reaction analysis, we propose a mechanism in which the first step of NO reduction is the adsorption of C_3H_6 on the catalyst surface, and a nitrogen-containing organic intermediate is generated from the adsorbed C_3H_6 and NO. A nitrogen molecule is produced in a process where the intermediate(carbonaceous material containing nitrogen) is decomposed by the reaction with another NO(or NO_2) molecule in the gas phase.

The reduction of nitrogen oxides(NOx) to molecular nitrogen is an important task for environmental chemistry. Recently, selective catalytic reduction(SCR) of NO by hydrocarbons in an oxygen-rich atmosphere has attracted considerable attention as a new type of reaction alternative for traditional NH_3-SCR process. This reaction was reported to proceed on various cation-exchanged zeolites (*1 - 5*), metallosilicates (*6, 7*) , Al_2O_3 (*8, 9*), and SiO_2-Al_2O_3 (*10*), when C_2H_4 or higher hydrocarbons were used as reductant. These studies showed that oxygen was a necessary component for this reaction system.

We have reported that Fe-silicate exhibited high catalytic activity for this reaction at a relatively low temperature when C_3H_6 is used as a reducing agent. We also showed that the carbonaceous material deposited on the catalyst surface was involved in this reaction (*6*). As for this type of selective reduction, several mechanisms have been proposed:
(1) Reaction between NO_2 and hydrocarbon (*11, 12*).
(2) Participation of an intermediate such as isocyanate, radicals, or oxygen-containing compound (*13-16*).
(3) Redox mechanism consisting of alternative oxidation and reduction of catalyst surface by NO and hydrocarbon (*17- 19*).
The mechanism, however, may vary from one type of catalyst to another, or it may depend on the kind of hydrocarbons used as reductant as well as the reaction conditions. In this study, we studied the reaction mechanism on Fe-silicate catalyst using a technique of transient response method.

---

[1]Corresponding author

0097–6156/95/0587–0123$12.00/0

Experimental

Catalyst Preparation   Fe-silicate(Si/Fe = 50, molecular ratio) used in this study was synthesized according to the rapid crystallization method reported by Inui et al.(20). Water glass(29wt% SiO$_2$, 9wt% Na$_2$O) and FeCl$_3$·6H$_2$O were used as the sources of silicon and iron, respectively. Tetrapropylammonium bromide (TPAB) was used as the template molecule. The mixture was heated in a stainless steel autoclave from room temperature to 433 K at a constant heating rate of 1.5 K·min^{-1}, and then up to 483 K at a constant heating rate of 12 K·h^{-1}.

The X-ray diffraction patterns of the synthesized Fe-silicate was similar to that of H-ZSM-5, indicating that the Fe-silicate had a pentasil pore-opening structure. Chemical compositions of synthesized Na-form Fe-silicate were determined by inductively coupled argon plasma atomic emission spectrometer (ICP) (21). The amount of sodium ion was in accordance with that of iron, showing that iron is mostly incorporated in the framework. Thus, the synthesized Fe-silicate was converted into H-form by the ion-exchange method using 1 M-NH$_4$NO$_3$ solution (353K for 1h, 5 times), followed by calcination at 813 K for 3 h. Sodium was hardly detected after this ion-exchange.

Transient response study   A mixture of 1000 - 2000 ppm of NO, 500 - 1000 ppm C$_3$H$_6$, and 10% O$_2$ was fed to 0.5 g of catalyst at a rate of 100 cm^3·min^{-1}. Effluent gases were analyzed by means of gas chromatography and chemiluminescence detection of NO and NO$_2$. The catalytic activity was evaluated by the level of NO conversion to N$_2$.

Transient response experiments were conducted using a fixed-bed reactor which was equipped with 2 lines of reactant stream, which could be switched by use of a 4-way valve to change immediately one reactant gas composition to another. All dead volume was minimized. Effluent gases were stored in a multiposition 8-way loop valve, and analyzed in turn by means of on line gas chromatography. The first reaction run before switching the reactant gas will be called "Reaction A" and the latter run "Reaction B".

During Reaction A, the concentration of C$_3$H$_6$, CO$_x$, and NO$_x$ in the outlet gas were measured by means of gas chromatography and chemiluminescence detection of NO$_x$, respectively. The amounts of adsorbed C$_3$H$_6$ and NO$_x$ on the catalyst were calculated from the material balance of reactant and product. After switching the reactant components for various types (Reaction B), the variation in the behavior in N$_2$ formation were compared.

Results and Discussion

Effect of Reducing gases on NO reduction   As shown in our previous study (1), H-Fe-silicate is a highly active and stable catalyst for the reduction of NO with C$_3$H$_6$ at 573 K, and the first step of this reaction is considered to be the adsorption of C$_3$H$_6$, followed by the formation of carbonaceous material on the catalyst.

Table 1 summarizes the temperature dependence of NO conversion with various reducing gases on H-Fe-silicate. The level of NO conversion markedly depended on the kind of reducing gases. CO and H$_2$ were not effective for this reaction, and they were preferentially consumed by the reaction with O$_2$. Reduction of NO proceeded when hydrocarbons were used as reductants, and the reduction reactivity decreased in the following order:

C$_3$H$_6$ > C$_2$H$_4$ > C$_3$H$_8$ >> C$_2$H$_6$ > CH$_4$.

Table I . Effect of reducing gases on the catalytic activity of Fe-silicate.

Temperature / K	NO conversion to N_2(Hydrocarbon conversion to CO_x) / %			
	473	573	673	773
CH_4	-	1.33(1.00)	4.04(17.3)	4.22(74.0)
C_2H_4	25.9(28.2)	36.6(82.2)	15.5(94.6)	5.82(94.6)
C_2H_6	-	9.36(13.1)	12.5(74.3)	5.85(98.3)
C_3H_6	-	54.6(83.4)	16.5(100)	5.15(100)
C_3H_8	10.1(1.5)	26.1(37.7)	12.9(81.3)	9.20(100)

Reducing gases, 1000 ppm (C_3H_8, C_3H_6, C_2H_6, C_2H_4, CH_4);

NO, 1000 ppm; O_2 10%; total flow rate, 100 $cm^3 \cdot min^{-1}$; catalyst weight, 0.5 g.

Mechanism of NO reduction by Propene   We studied the reaction mechanism of NO reduction by C_3H_6 on Fe-silicate catalyst by means of transient response method in the following manner. First, the reaction was carried out using a mixture of NO, C_3H_6, and O_2 at 573 K (Reaction A). After the N_2 formation rate reached the steady-state, the reactant mixture was switched for various components.

Figure 1 shows the variation in the behavior of N_2 formation when the components of the reactant mixture in Reaction B were switched for various types. When only NO or O_2 in He alone was introduced, or when a mixture of C_3H_6 and O_2 was introduced, N_2 was hardly produced. On the other hand, when a mixture of NO and O_2 was introduced, a large amount of N_2 was produced.

Carbonaceous material was deposited on the catalyst surface simultaneously (*6*) with the formation of N_2 and CO_x in Reaction A (Figure 2) . After a mixture of NO and O_2 was introduced(Reaction B), CO_x was produced simultaneously with the formation of N_2. However, the formation of N_2 decreased gradually and completely terminated after 60 min with no more formation of CO_x. From these results, it is suggested that carbonaceous material deposited on the catalyst surface plays an important role in this reaction and that the first step of the reaction is the adsorption of C_3H_6 on the catalyst surface.

As shown in Fig.3-(a), when the reactant component was switched from the mixture of NO, C_3H_6, and O_2 into the mixture of NO and O_2, the ln(N_2 formation rate) and reaction time at early stage in Reaction B have a linear relationship. In addition to this, when N_2 formation rate is extrapolated to the moment of switching the reaction component, the rate coincides with the steady state N_2 formation rate of Reaction A. Therefore, from these results, it can be concluded that N_2 is produced by the reaction between an intermediate and NO + O_2, and first order reaction with respect to the concentration of reaction intermediate.

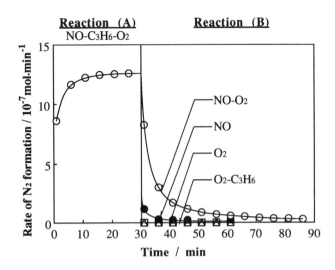

Fig.1   Effect of reactant components on NO reduction to N$_2$.
Reaction A: C$_3$H$_6$, 1000 ppm; NO, 1000 ppm; O$_2$, 10%.
Reaction B: O: NO, 1000 ppm; O$_2$, 10%.
          Δ: O$_2$, 10%; C$_3$H$_6$, 1000 ppm .
          □: O$_2$, 10%. ●: NO, 1000 ppm.

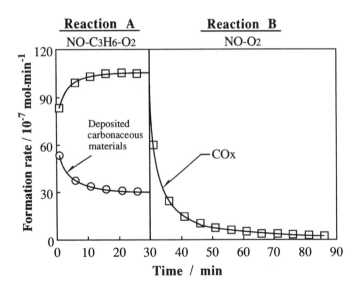

Fig.2   Time-on-stream variation in the formation rate of
carbonaceous material and COx.
Reaction A: C$_3$H$_6$, 1000 ppm; NO, 1000 ppm; O$_2$, 10%.
Reaction B: NO, 1000 ppm; O$_2$, 10%.

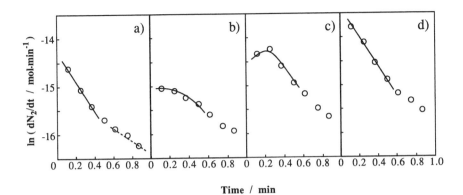

Fig.3 Relationship between ln(dN$_2$/dt) and reaction time
in Reaction B when reactant components was changed.
Reaction A: (a): NO, 2000 ppm; C$_3$H$_6$, 1000 ppm; O$_2$, 10%.
(b): C$_3$H$_6$, 1000 ppm, O$_2$ 10%. (c): C$_3$H$_6$, 1000 ppm;
(d): NO, 2000 ppm; C$_3$H$_6$, 1000 ppm.
Reaction B: NO, 2000 ppm; O$_2$, 10%.

In order to obtain further insight into the carbonaceous material, the effects of the reactant components of Reaction A on the behavior of N$_2$ formation in Reaction B were studied.

First, when the mixture of C$_3$H$_6$ and O$_2$ in He was introduced in Reaction A, the rate of N$_2$ formation in Reaction B was smaller than that of the steady-state reaction (Figure 3-(b)). Therefore, the carbonaceous material formed by the reaction between C$_3$H$_6$ and O$_2$ also promotes the reaction. However, the reactive intermediate could not be formed. Probably, the carbonaceous material formed by the reaction between C$_3$H$_6$ and O$_2$ should also contribute to N$_2$ formation at later stage of the Reaction B (shown in Fig.3-(a): dotted line), however, this is not the main pathway of NO reduction.

Second, when only C$_3$H$_6$ in He was introduced in Reaction A, the rate of N$_2$ formation in Reaction B was almost the same as with the C$_3$H$_6$, NO, O$_2$ mixture. However, the induction period was observed at an early stage (Figure 3-(c)). Therefore, adsorption of C$_3$H$_6$ on the catalyst surface should be the first step of this reaction. However, the C$_3$H$_6$ adsorbed on the catalyst could not form the reactive intermediate.

Finally, when a mixture of NO and C$_3$H$_6$ was introduced in Reaction A, the rate of N$_2$ formation in Reaction B was the same as during steady-state reaction. Therefore, coexistence of NO and C$_3$H$_6$ is necessary to form the reaction intermediate.

Figure 4-(a) shows the variation in the formation rate of the deposited carbonaceous material in Reaction A, determined by the material balance, whereas Fig.4-(b) shows the variation in the formation rate of N$_2$.

After one of the gas mixtures of NO + C$_3$H$_6$ + O$_2$, C$_3$H$_6$ + O$_2$, NO + C$_3$H$_6$, C$_3$H$_6$ was fed for 30 minutes, followed by the deposition of carbonaceous material on the catalyst, the reactant gas was switched to NO and O$_2$. The amount of the deposited carbonaceous materials decreased in the order:

$$NO + C_3H_6 + O_2 > C_3H_6 + O_2 > NO + C_3H_6 \approx C_3H_6$$

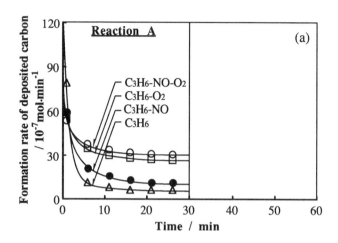

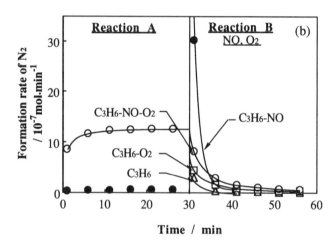

Fig.4   Effect of reactant components on the formation of
carbonaceous materials and N$_2$.
(a) Formation rate of deposited carbon in Reaction A.
(b) Formation rate of N$_2$.
Reaction A: O: C$_3$H$_6$, 1000 ppm; NO, 1000 ppm; O$_2$, 10%.
              □: C$_3$H$_6$, 1000 ppm; O$_2$, 10%.
              △: C$_3$H$_6$, 1000 ppm.
              ●: C$_3$H$_6$, 1000 ppm; NO, 1000 ppm.
Reaction B: NO, 1000 ppm; O$_2$, 10%.

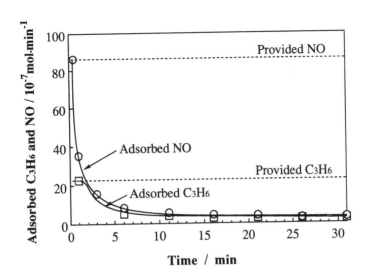

Fig.5 Time-on-stream variation in the rate of adsorbed C_3H_6 and NO
in NO-C_3H_6 reaction.
O, adsorbed NO; □, adsorbed C_3H_6.

In the case of NO + C_3H_6 treatment, N_2 formation started immediately after switching feed stream to NO + O_2 in spite of small amount of deposited carbonaceous material. The carbonaceous material deposited upon treatment with C_3H_6 + NO is extremely reactive, showing that carbonaceous species effective for NO reduction is probably produced by the coexistence of C_3H_6 and NO.

Also when reactant components were switched in the order of NO → C_3H_6 → NO+O_2, or C_3H_6 → NO → NO+O_2, the behavior of N_2 formation was almost same with the order of C_3H_6 → NO+O_2 in this reaction conditions.

As shown in Fig.5, when a mixture of NO and C_3H_6 was introduced in Reaction A, both NO and C_3H_6 were adsorbed on the catalyst surface, and the adsorbed molar ratio of NO to C_3H_6 was unity.

From the results that the coexistence of C_3H_6 and NO was necessary to form the reactive intermediate, the intermediate is suggested to be produced by the reaction between the unstable carbon species such as radical or π-aryl and NO. Thus, it is suggested that NO reduction proceeds via decomposition of the nitrogen containing organic compounds.

Figure 6 shows the effect of the amount of reaction intermediate on N_2 formation in Reaction B. The reaction time in Reaction A was varied from 0.5 min to 30 min in order to change the amount of reactive intermediate. At 0.5 min and 1 min, the N_2 formation rate in Reaction B was same. On the other hand, when NO and C_3H_6 were introduced for 30 min in Reaction A, a limit on N_2 formation was observed in the

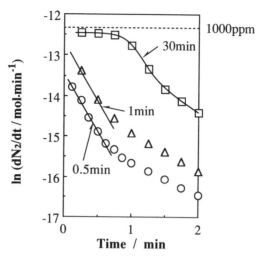

Fig.6 Effect of the amount of carbonaceous material formed
from C_3H_6 and NO on the rate of N_2 formation.
Reaction A: C_3H_6, 2000 ppm; NO, 500 ppm.
　　　　　Reaction time: O, 0.5 min; △, 1 min; □, 30 min.
Reaction B: NO, 1000 ppm; O_2, 10%.

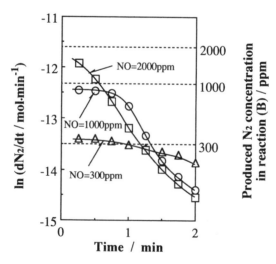

Fig.7 Effect of NO concentration in reaction (B) on the rate
of N_2 formation.
Reaction A: C_3H_6, 2000 ppm; NO, 500 ppm. Reaction time, 30 min.
Reaction B: O: NO, 300 ppm; O_2, 10%.
　　　　　□: NO, 1000 ppm; O_2, 10%.
　　　　　△: NO, 2000 ppm; O_2, 10%.

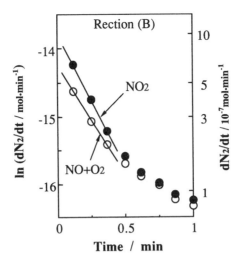

Fig.8 Effect of NO_2 on the rate of N_2 formation in reaction (B).
Reaction A: NO, 2000 ppm; C_3H_6, 500 ppm; O_2, 10 %; 0.5 min
Reaction B:O; NO, 1000 ppm; O_2, 10 %
●; NO_2, 1000 ppm

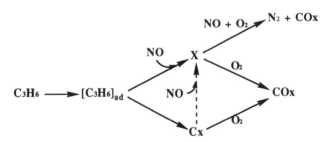

Scheme 1   Proposed reaction mechanism of NO reduction on Fe-silicate catalyst.

early stage in Reaction B. As shown in Fig.7, N_2 formation in Reaction B was limited by the NO concentration in Reaction B. For instance, when 1000 ppm of NO was introduced, the maximum N_2 produced was 1000 ppm. Therefore, N_2 should be produced by the reaction between a nitrogen-containing reaction intermediate and a gas phase NO(or NO_2) molecule.

As mentioned above, the coexistence of NO and O_2 was necessary to produce N_2 in Reaction B. It has been reported that NO_2 plays an important role in NO reduction by hydrocarbons (*12, 22*). Therefore, the effect of NO_2 on N_2 formation rate in Reaction B was studied. As shown in Figure 8, when NO_2 in He alone was introduced in Reaction B, N_2 formation rate was greater than when NO and O_2 mixture was introduced. This suggests that NO_2 plays an important role in the decomposition of the nitrogen-containing organic compound.

Conclusion

From these results, we propose a reaction mechanism in which the first step of NO reduction is the adsorption of C_3H_6 on the catalyst surface, and a nitrogen-containing organic intermediate is formed by the reaction between the adsorbed C_3H_6 and NO (Scheme 1). A nitrogen molecule seems to be produced in a process where the carbonaceous material containing nitrogen (intermediate) is decomposed by the reaction with another NO(or NO$_2$) molecule in the gas phase.

Literature Cited

1) S. Sato, Y. Yu-u, H. Yahiro, N. Mizuno, and M. Iwamoto, *Appl.Catal.*, **70**, L1 (1991).

2) H. Hamada, Y. Kintaichi, M. Sasaki, T. Itoh, and M. Tabata, *Appl. Catal.*, **64**, L1 (1990).

3) S. Sato, H. Hirabayashi, H. Yahiro, N. Mizuno, and M. Iwamoto, *Catal. Lett.*, **12**, 193 (1992).

4) M. Misono, and K. Kondo, *Chem. Lett.*, **1991**, 1001.

5) K. Yogo, S. Tanaka, M. Ihara, T. Hishiki, and E. Kikuchi, *Chem. Lett.*, **1992**, 1025.

6) E. Kikuchi, K. Yogo, S. Tanaka, and M. Abe, *Chem. Lett.*, **1991**, 1063.

7) T. Inui, S. Iwamoto, S. Kojo, and T. Yoshida, *Catal. Lett.*, **13**, 87 (1992).

8) Y. Kintaichi, H. Hamada, M. Tabata, M. Sasaki, and T. Ito, *Catal. Lett.*, **6**, 239, (1990).

9) Y. Torikai, H. Yahiro, N. Mizuno, and M. Iwamoto, *Catal. Lett.*, **9**, 91 (1992).

10) H. Hosose, H. Yahiro, N. Mizuno, and M. Iwamoto, *Chem. Lett.*, **1991**, 1859.

11) J. O. Petunchi, G. Sill, and W. K. Hall, *Appl. Catal. B.*, **2**, 303 (1993).

12) M. Sasaki, H. Hamada, Y. Kintaichi, and T. Ito, *Catal. Lett.*, **15**, 297 (1992).

13) M. Iwamoto, H. Yahiro, and N. Mizuno, *Proc. 9th Int. Zeolite Conf.*, Vol II, 397 (1992).

14) Y. Ukisu, S.Sato, G. Muramatsu, and T. Yoshida, *Catal., Lett.*, **11**, 177 (1991).

15) A. Obuchi, A. Ogata, K. Mizuno, A. Ohi, M. Nakamura, and H. Ohuchi, *J. Chem. Soc., Chem. Commun.*, 247 (**1992**).

16) A. Obuchi, A. Ogata, K. Mizuno, A. Ohi, M. Nakamura, and H. Ohuchi, *SHOKUBAI(CATALYST)*, **34**, 360 (1992).

17) T. Inui, S. Kojo, T. Yoshida, M. Shibata, and S. Iwamoto, *SHOKUBAI(CATALYST)*, **33**, 77 (1991).

18) T. Inui, S. Kojo, M. Shibata, T. Yoshida, and S. Iwamoto, Stud. Surf. Sci. Catal., **69**, 355 (1991).

19) T. Inui, S.Iwamoto, S. Kojo, and T. Yoshida, *Catal. Lett.*, **13**, 87 (1992).

20) T. Inui, O. Yamase, K. Fukuda, A. Itoh, J. Tarumoto, N. Morinaga, T. Hagiwara, and Y. Takegami, *Proc. 8th Intern. Cong. Catal.*, Berlin, 1984, III-p.569.

21) K. Yogo, S. Tanaka, T. Ono, T. Mikami, and E. Kikuchi, *Microporous Materials*, in press.

22) K. Yogo, M. Umeno, H. Watanabe, and E. Kikuchi, *Catal. Lett.*, **19**, 131 (1993).

RECEIVED November 16, 1994

Chapter 11

Decomposition of NO over Metal-Modified Cu-ZSM-5 Catalysts

Yanping Zhang[1], Tao Sun[2], Adel F. Sarofim[1], and Maria Flytzani-Stephanopoulos[2]

[1]Department of Chemical Engineering, Massachusetts Institute of Technology, Cambridge, MA 02139
[2]Department of Chemical Engineering, Tufts University, Medford, MA 02155

Alkaline earth and rare earth metal cocation effects are reported in this paper for copper ion-exchanged ZSM-5 zeolites used for the catalytic decomposition of nitric oxide in O_2- free, O_2- rich, and wet streams. Severe steaming (20% H_2O) of Na-ZSM-5 at temperatures above 600°C leads to partial vitreous glass formation and dealumination. Unpromoted Cu-ZSM-5 catalysts suffer drastic loss of NO decomposition activity in wet gas streams at 500°C. Activity is partially recovered in dry gas. Copper migration out of the zeolite channels leading to CuO formation has been identified by STEM/EDX. In Ce/Cu-ZSM-5 catalysts the wet gas activity is greatly improved. CuO particle formation is less extensive and the dry gas activity is largely recovered upon removal of the water vapor.

The removal of nitric oxide from exhaust gas streams in power plants, industrial boilers and engine systems continues to be a challenge in view of new more stringent regulations. A particularly high-priority need is the efficient control of NO_x in lean-burn exhausts. The simplest method for NO_x control would be catalytic decomposition to benign N_2 and O_2. Iwamoto and coworkers (1) first reported that Cu^{2+}-ZSM-5 zeolites have stable activity for the direct decomposition of NO even in the presence of oxygen. Subsequently several studies of this catalyst system followed (2-5). The main problems associated with further development of the Cu-ZSM-5 catalyst for the NO decomposition are reduced activity at high-temperatures (>500°C), hydrothermal stability and poisoning by SO_2. The latter problem can be handled in many exhausts by separate removal of SO_2 upstream of the NO_x catalyst.

Kagawa and coworkers (6) recently reported that the incorporation of cocations into Cu-ZSM-5 catalysts promoted the high-temperature activity (above 450°C) for NO decomposition in O_2-free gas streams. Work in this laboratory has further shown a positive cocation effect on the NO decomposition in both O_2-free and O_2-containing gases with various NO concentrations (7). For the positive effect to be displayed, it is necessary to follow a specific mode of ion exchange during catalyst preparation.

0097–6156/95/0587–0133$12.00/0

All NO$_x$-containing combustion gases also contain a significant amount of water vapor (2-15%), therefore resistance to water poisoning is important for practical application of a catalyst. Iwamoto, *et al* (8), and Li and Hall (9) reported that the catalytic activity of Cu-ZSM-5 for NO conversion to N$_2$ was decreased in the presence of 2% water vapor, but it could be recovered after removal of water vapor. However, Kharas and coworkers (10) found that Cu-ZSM-5 was severely deactivated under a typical automotive fuel lean exhaust gas (10% H$_2$O, GHSV = 127,000 h^{-1}) over the temperature range of 600 to 800°C. The authors attributed the catalytic deactivation of Cu-ZSM-5 to the formation of CuO and the disruption of zeolitic crystallinity and porosity.

In the present study, we address effects of cocations on the catalytic activity of Cu-ZSM-5. Metal modified Cu-ZSM-5 catalysts are tested in both O$_2$-free and O$_2$-containing gases, as well as in wet (2-20% H$_2$O) gas streams.

Experimental

Catalyst Preparation. Catalysts were prepared by incorporating metal cations into Na-ZSM-5 zeolites according to well-established ion exchange procedures. Nitrates of the cocations and cupric acetate aqueous solutions (0.007 M) were used. The sequence of exchange was cocation first, followed by drying and calcining, then copper ion exchange as reported before (7). The starting materials were either the as-received Na-ZSM-5 zeolite (SMR 6-2670-1191, Davison Chemical Division, W. R. Grace & Co.) with Si/Al = 21.5 or steamed Na-ZSM-5. In catalysts containing Ce^{3+}, the ZSM-5 zeolite was ion-exchanged with the cerous nitrate hexahydrate (Fluka, assay> 99.0%) in dilute aqueous solution with concentration of 0.007 M. The exchange was made at 85°C for 2 hours. After filtration, the Ce ion-exchanged ZSM-5 zeolite was dried at 100°C for 10 hours, and further calcined in a muffle furnace in air at 500°C for 2 hours. The reason for calcining the materials was to stabilize Ce in the zeolite. The Ce ion-exchange with the ZSM-5 was performed thrice. Very small Ce clusters on the external surface of ZSM-5 and uniformly distributed Ce were identified by STEM/EDX. It is proposed that the Ce clusters are in oxide form, and the uniformly distributed Ce are associated with zeolite framework aluminum. As shown in Table 1, the Na/Al ratio is reduced confirming ion exchange. Elemental analyses were performed by Inductively Coupled Plasma Emission Spectrometry (ICP, Perkin-Elmer Plasma 40) after the catalysts were dissolved in reagents purchased from UniSol for measuring the Ce concentration, since Ce^{3+} cations are fluoride insoluble. These reagents comprise three types of solutions. One, containing HF, dissolves the catalyst, the other two neutralize and stabilize the solution to (a) deactivate the HF by increasing the pH to a value of 7.5 to 8.0, and (b) maintain solubility of the sample. In the text, the catalysts are identified in the following way: M(percent cocation exchange level)/Cu^{2+}(percent exchange level)-ZSM-5, where 100% metal ion (of n valence) exchange level in the catalysts is defined as one metal ion replacing n Na$^+$ [or neutralizing n (AlO$_4$)$^-$] and the atomic ratio of (metal ion)/Al = 1/n, e.g., Cu^{2+}/Al = 0.5. The catalysts prepared and tested in this work are shown in Table 1.

Reaction Measurements. The NO decomposition activity of the catalysts was evaluated in a laboratory-scale reactor system described in detail elsewhere (7). An amount of 0.5 g of catalyst was placed in the reactor for NO conversion measurements. The contact time, W/F, defined as the ratio of catalyst weight in the reactor to the total flowrate of the feed gas stream, was 1.0 g s/cm^3 (STP). A gas mixture of 2% NO, 0-5% O$_2$, 0-20% H$_2$O and balance He was used in the tests.

Hydrothermal effects on the structure and Cu^{2+} ion exchange capacity of the parent Na-ZSM-5 zeolites, and on the catalytic activity of the Cu-ZSM-5 and metal ion modified Cu-ZSM-5 catalysts for NO decomposition were evaluated. The parent Na-ZSM-5 zeolite was pretreated for 20 hours in a gas mixture containing 20% H_2O- 4% O_2- He at temperatures of 500, 600 and 750°C to examine zeolite structural changes. Micropore volumes of the as-received and the steamed Na-ZSM-5 zeolites were measured by nitrogen uptake in a Micropore Analyzer (Micromeritics, ASAP 2000), as well as by n-hexane adsorption in a TGA at 25°C. The amount of n-hexane adsorbed was obtained after the sample weight no longer increased with time in a n-hexane and nitrogen stream. Assuming liquid n-hexane (density: 0.62 g/cc) in the pores, the volume of n-hexane is equal to the sample micropore volume. No correction for surface adsorbed n-hexane was made here *(11)*. The Cu^{2+} uptake capacity of the steamed samples was examined by the usual Cu^{2+} ion exchange procedure. Comparison of wet-gas catalytic activities of Cu-ZSM-5 with those of metal modified Cu-ZSM-5 samples for NO decomposition were made in a feed gas mixture of 2- 20% H_2O- 2% NO- balance He at 500°C. Nitric oxide conversions to nitrogen were measured before, during and after introduction of water vapor into the feed stream. All measurements were made after the reaction had reached a steady-state.

Characterization of Catalyst Samples. A Rigaku 300 X-Ray powder diffractometer was used to examine the crystal structures of the fresh and steamed catalysts. The samples were evacuated at 55°C for 24 hours, then placed in a dessiccator with a drying agent. The diffraction patterns were taken in the 2θ ranges of 5- 80° at a scanning speed of 1° per minute. A VG HB5 Scanning Transmission Electron Microscope was used to characterize the fresh and steamed catalyst samples. The catalyst samples were supported by a 200 mesh, carbon coated, plain nickel grid (Ernest F. Fullam Inc. Part No. 14581) in the STEM chamber. Prior to placing the fresh samples on the nickel grids, they were calcined in a muffle furnace in air at 500°C for 2 hours. The calcined samples on the grids were further coated by carbon in vacuum to ensure no particle charging during the STEM analysis. The composition of interested area on catalyst samples was analyzed by Energy Dispersion X-Ray (EDX) microprobe technique (spot size: 0.5 nm by 1.0 nm).

Results and Discussion

Cocation Effect on Catalytic Activity in Dry NO-Containing Gases. Two sets of catalysts, namely Cu(97)-ZSM-5 and Mg(34)/Cu(86)-ZSM-5; Cu(141)-ZSM-5 and Ce(60)/Cu(138)-ZSM-5, respectively, with *ca* 90% and 140% Cu exchange levels were evaluated in a gas mixture of 2% NO -He, at a contact time of 1.0g s/cm^3 (STP) over the temperature range of 350-600°C. We have previously reported *(7)* that for the same Cu^{2+} ion-exchange level (*ca* 70%), the Mg(52)/Cu(66)-ZSM-5 catalyst showed significantly higher NO conversion to N_2 than Cu(77)-ZSM-5 in the high temperature region (450- 600°C), while a low Ce^{3+}-exchanged catalyst, Ce(11)/Cu(119)-ZSM-5, promoted the activity of Cu-ZSM-5 in the low temperature region (300- 450°C). In this study, the Cu exchange levels of both Mg/Cu- and Ce/Cu-ZSM-5 catalysts are higher than in the previously reported catalysts *(7)*. The cocation ions still have positive effect on the catalytic activity. However, the relative increment of NO conversion to N_2, although the absolute catalytic activities are higher, is smaller than that over the low Cu ion-exchanged Mg/Cu- and Ce/Cu-ZSM-5. This is shown in Figures 1 and 2 in terms of conversions of NO to N_2 over the Cu(97)-ZSM-5 and Mg(34)/Cu(86)-ZSM-5, and the Cu(141)-ZSM-5 and Ce(60)/Cu(138)-ZSM-5 catalysts. It should be pointed out that the Mg-ZSM-5 and Ce-ZSM-5 materials have, respectively, zero and negligible (<10%)

Table 1. Summary of Catalyst Preparation[1]

Catalyst (M/Cu-ZSM-5)	Si/Al[2]	Cu/Al[2]	M/Al[2]	Na/Al[2]	Cu exchange
as-received	21.5[3]			1.0	
Cu	21.9	0.435 (97%)		0.18	once, RT
Cu	20.3	0.705 (141%)		0.0	thrice, RT
Mg/Cu	17.1	0.43 (86%)	0.17 (34%)	~0	twice, RT
Ba/Cu	22.1	0.63 (126%)	0.025 (5%)	~0	twice, RT
Y/Cu	22.4	(0.675) (135%)	0.045 (13%)	~0	twice, RT twice, RT
Ce/Cu	19.5	0.596 (119%)	0.036 (11%)	~0 (0.58)[4]	twice, RT
Ce/Cu	19.8	0.69 (138%)	0.20 (60%)	~0 (0.52)[4]	thrice, RT

[1] Cocation exchanged once, except Ce(60)/Cu(138)-ZSM-5 thrice, at 85°C for 2 hours.

[2] Measured by ICP; the values in parentheses are ion exchange levels calculated from M/Al ratios by multiplying by the assumed valence x 100%.

[3] As-received Na-ZSM-5: SMR-2670-1191 (Davison).

[4] the values in parentheses are Na/Al ratios for the Ce-ZSM-5 materials, prior to copper exchange.

Table 2. Hydrothermal Effects[1]

Catalyst	as-received Na-ZSM-5[2]	Steamed Na-ZSM-5		
Steaming Temperature (°C)	-	500	600	750
Micropore Volume (cm^3/g)	0.11[3] (0.147)[4]	0.11[3] (0.109)[4]	0.0[3] (0.0)[4]	NM NM
Cu Exchange Capacity (%)	115[5]	104[5]	20[6]	15[6]

[1] in 20% H$_2$O- 4% O$_2$- 76% He; 20 h.

[2] Davison Chemical Division, W. R. Grace. Co., Si/Al = 21.5.

[3] Measured by N$_2$ uptake in a Micropore Analyzer.

[4] Measured by n-hexane adsorption.

[5] In two ion exchange steps, from 0.007 M Cu(ac)$_2$ aqueous solution, 20 hours, at room temperature.

[6] In three ion exchange steps, from 0.007 M Cu(ac)$_2$ aqueous solution, 20 hours, at room temperature.

NM: not measured.

activity for NO decomposition. However, the corresponding cocation modified Cu-ZSM-5 catalysts displayed significant promotion effect in 5% O_2-containing gas, as shown in Figures 1 and 2.

The cocation promotion effect in O_2-free gas streams can be explained by the hypothesis that the cocations occupy "hidden" sites for NO decomposition, thus, driving the later exchanged Cu^{2+} cations preferentially onto the active sites (7). This implies that the positive cocation effect should be more pronounced for low Cu exchange levels, since the cocations are inert for the reaction, and a limit of maximum cation exchange capacity exists in ZSM-5 . Earlier obtained results support the above hypothesis: (i) catalysts prepared by incorporating Cu^{2+} first, and cocation second did not show the cocation promotion effect for NO decomposition (6, 7), (ii) the performance of metal ion modified Cu-ZSM-5 was not sensitive to the types of cocations chosen. Thus, the NO to N_2 conversion profiles over several alkaline earth and transition metal ion modified Cu-ZSM-5 catalysts with a similar Cu^{2+} exchange level overlapped over the whole temperature range of 350-600°C (7). In the presence of oxygen, however, a mere structural promotion effect does not appear adequate to explain the data. An electronic effect is more plausible here. This may be understood as preserving the Cu^+/Cu^{2+} ratio in the range active for NO decomposition (2).

Structural Effects of Water Vapor on Na-ZSM-5. The catalytic activity of Cu-ZSM-5 and Mg/Cu-ZSM-5 catalysts was permanently lost after a gas mixture of 20% H_2O-4%O_2-He had flowed through the catalyst bed at 750°C for 20 hours (12). The loss of catalytic activity may be attributed to either dealumination of the ZSM-5 material or deactivation of copper, or both. To check for dealumination, the parent Na-ZSM-5 zeolite was pretreated for 20 hours in a gas mixture containing 20% H_2O- 4% O_2-He at 500, 600, and 750°C. It was found that the micropore volume decreased from 0.11 cm^3/g for the as-received Na-ZSM-5 to practically zero after treatment at 600°C, while the subsequent Cu^{2+} ion exchange capacity was reduced from 141% to 20% (or 15% for the sample steamed at 750°C). These Cu-ZSM-5 catalysts had very low NO decomposition activity. These results are shown in Table 2.

The effects of steaming were further examined by XRD analysis of the catalysts. Figure 3 shows the diffractograms of the as-received, and the steamed Na-ZSM-5 catalysts at 500 and 600°C. Except for very fine particle appearance indicated by small angle scattering, no significant structure change of the sample steamed at 500°C is observed by comparing its XRD pattern with that of the as-received Na-ZSM-5. However, after steaming the Na-ZSM-5 at 600°C vitreous glass formation (13) is indicated by the appearance of an amorphous background (halo) on the diffraction pattern at $2\theta = $ 10- 40°. The ensuing very low copper cation exchange capacity, and micropore volume loss (Table 2) together with Figure 3 provide evidence of a major structural change for the 600°C-treated Na-ZSM-5. This severe steaming can also lead to dealumination, which is best indicated by the reduced ion-exchange uptake capacity, since cations in the Na-ZSM-5 are associated with $(AlO_4)^-$. On the other hand, little micropore volume loss and little Cu^{2+} uptake capacity loss was found for the 500°C-steamed Na-ZSM-5. This fact suggests that the 500°C-steamed Na-ZSM-5 undergoes less significant structural change, compared with the 600°C-steamed Na-ZSM-5. Therefore, it is at this temperature of 500°C that the effect of water vapor on the catalytic activity for all Cu-ZSM-5 and metal modified Cu-ZSM-5 materials was examined to correlate it with the stability of copper in the presence of only small structural modifications of the zeolite.

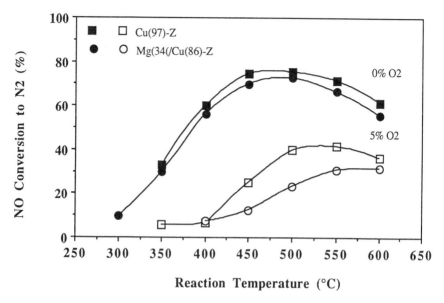

Figure 1.   Activity comparison of Mg(34)/Cu(86)-ZSM-5 with
             Cu(97)/ZSM-5 in 2% NO-He, at W/F = 1g s/cm^3 (STP), and 0% and
             5% O$_2$.

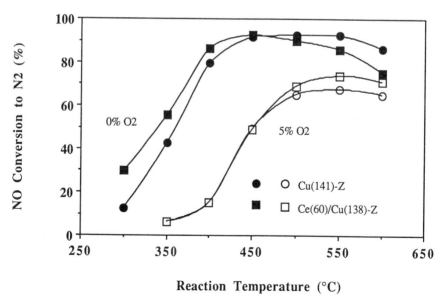

Figure 2.   Activity comparison of Ce(60)/Cu(138)-ZSM-5 with Cu(141)-ZSM-5
             in 2% NO- He, at W/F=1 g s/cm^3, and 0% and 5% O$_2$.

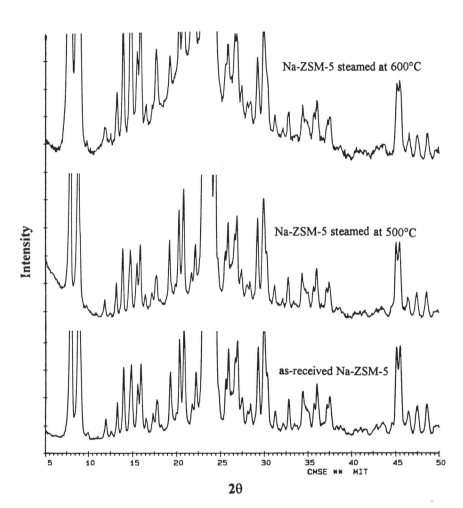

Figure 3.   XRD patterns of Na-ZSM-5 samples reveal evolution of vitreous glass formation with steaming temperature (in 20% H_2O-4% O_2-76% He for 20 hours).

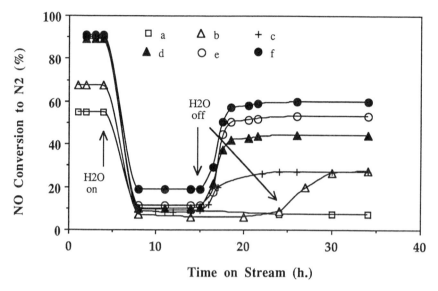

Figure 4.   Effect of H$_2$O vapor on the NO conversion to N$_2$ at 500°C over (a)
Cu(72)-ZSM-5; (b) Mg(52)/Cu(66)-ZSM-5; (c) Cu(141)-ZSM-5; (d)
Ba(5)/Cu(126)-ZSM-5; (e) Y(13)/Cu(135)-ZSM-5; (f)
Ce(60)/Cu(138)-ZSM-5 catalysts in 20% H$_2$O- 2% NO-He, W/F=
1g s/cm^3 (STP).

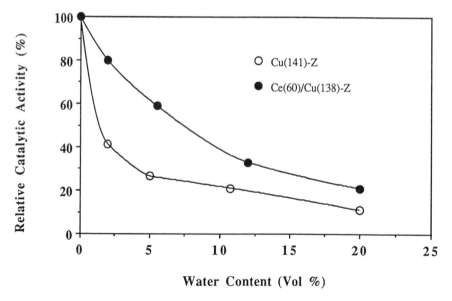

Figure 5.   Wet gas activity comparison of Ce(60)/Cu(138)-ZSM-5 with
Cu(141)-ZSM-5 in 2% NO-He, W/F= 1g s/cm^3 (STP) and
500°C.

Cocation Effect in Wet NO-Decomposition. A series of tests was performed in 2% NO-He at contact time 1g s/cm^3 and 500°C under dry (0% H$_2$O) as well as wet (2-20% H$_2$O) conditions over Cu-ZSM-5 and metal modified Cu-ZSM-5 catalysts. Figure 4 shows the steady-state NO conversion to N$_2$ over the catalysts Cu(72)-, Mg(52)/Cu(66)-, Cu(141)-, Ba(5)/Cu(126)-, Y(13)/Cu(135)- and Ce(60)/Cu(138)-ZSM-5 in dry gas (2% NO- He) and the wet gas mixture of 20% H$_2$O- 2% NO- He. Introduction of water vapor into the reactant gas mixture after the catalysts had been in dry NO-gas for 4 hours drastically decreased the conversion. The wet-gas activities reached a steady state corresponding to 8% NO conversion to N$_2$ for all catalysts, except 20% conversion for Ce(60)/Cu(138)-ZSM-5, in two hours.

The activity recovery in dry NO-gas after removal of water vapor is shown in Figure 4. The Mg(52)/Cu(66)-ZSM-5 gradually restored some of its activity to *ca* 45% of its original value. However, no recovery was found for the Cu(72)-ZSM-5 catalyst. For the over- exchanged Cu(141)-ZSM-5, 30% of its original dry gas activity was recovered after removal of water vapor. The cocation (Mg^{2+}, Ba^{2+}, Y^{3+} and Ce^{3+}) modified Cu-ZSM-5 catalysts recovered a larger fraction of their original catalytic activity than Cu(141)-ZSM-5. Among these cations, Ce^{3+} cations displayed the most pronounced positive effect on the wet gas-Cu^{2+} activity as well as on dry gas-activity recovery (more than 66%). The Ce/Cu-ZSM-5 catalyst was, thus, chosen for additional testing.

Activities of the catalysts Ce(60)/Cu(138)-ZSM-5 and Cu(141)-ZSM-5 were studied in a 2% NO- He stream with varying water vapor contents from 2 to 20% at 500°C. The wet gas activities were measured 8 hours after the NO conversion reached steady state. Relative catalyst activities to dry gas-activity are shown in Figure 5. A much faster deactivation is observed for Cu(141)-ZSM-5 than Ce(60)/Cu(138)-ZSM-5 upon exposing to water vapor. The latter catalyst displays twice as high activity throughout the water vapor content range of 2- 20%. Together with data shown in Figure 4, these results indicate that cocations, especially Ce^{3+}, can preserve the copper activity better in wet NO decomposition, and restore to a larger extent the activity of the Cu active sites after the removal of water vapor from the gas stream.

X-ray Diffraction Patterns. XRD patterns of the fresh and used Cu(141)-ZSM-5 and Ce(60)/Cu(138)-ZSM-5 catalysts show no significant loss of crystallinity. The diffractograms of the catalysts used in wet NO decomposition as well as the 500°C-steamed Na-ZSM-5 are shown in Figure 6. No large difference are apparent. XRD analysis of the highly siliceous ZSM-5 cannot exclusively determine whether the zeolite is dealuminated or not. Further examination of this issue is currently underway.

Characterization of Ce(60)/Cu(138)-ZSM-5 and Cu(141)-ZSM-5 by STEM. Electron micrographs representative for the Cu(141)-ZSM-5 catalyst in the freshly calcined state and after wet NO-decomposition are shown in Figure 7. The phase contrast imaging demonstrated formation of particles, which were also identified by EDX, clearly distinguishable from the support. Few copper oxide aggregates on the exterior surface of the freshly calcined Cu(141)-ZSM-5 are found, as shown in Figure 7a. The steamed Cu(141)-ZSM-5 catalyst, however, shows many CuO particles with mean aggregate sizes of 4- 6 nm (bright spots on dark field image as seen in the Figure 7b). On the other hand, fewer particles on the surface and uniform Cu distribution are shown in Figure 8 for the steamed Ce(60)/Cu(138)-ZSM-5. Detailed characterization of the catalysts by STEM/EDX mappings and XRD diffraction patterns to be reported in another publication are in good agreement with the above observations. There was no indication for sintering caused by the focused STEM electron beam. Active copper loss and activity recovery pathways are proposed in the following simplified scheme:

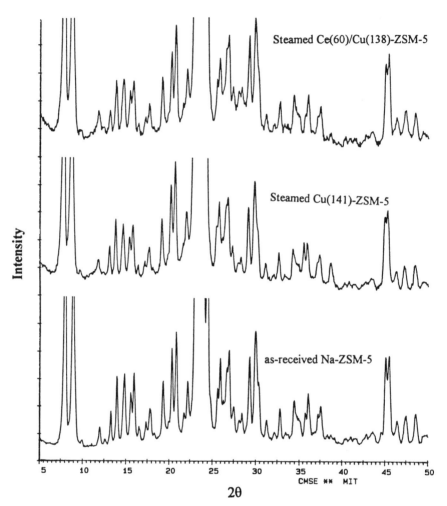

Figure 6.   XRD patterns of the steamed Na-ZSM-5 (20 hours) in 20%
             H$_2$O- 4% O$_2$- He, and Cu(141)-ZSM-5 and Ce(60)/Cu(138)-
             ZSM-5 (10 hours) in 20% H$_2$O- 2% NO- He at 500°C.

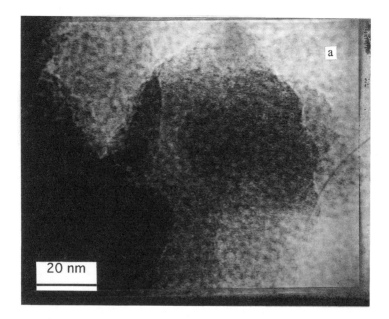

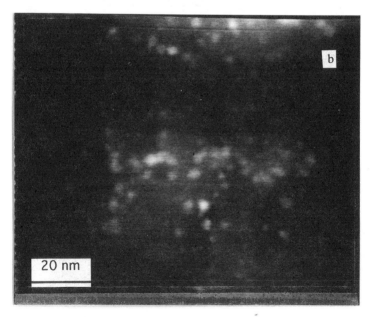

Figure 7.   Scanning transmission electron micrographs of the Cu(141)-ZSM-5:
(a) fresh sample calcined in air at 500°C for 2 hours; (b) copper
aggregates (bright spots) on the sample reacted at 500°C in 20%
H_2O-2% NO-He for 10 hours, then in dry gas (2% NO-He) for 20
hours.

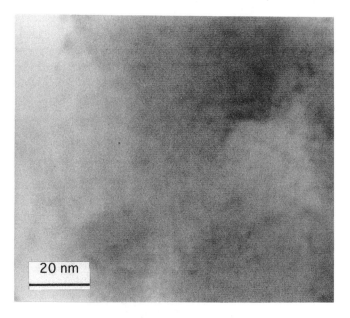

Figure 8.   Scanning transmission electron micrograph of the
                  Ce(60)/Cu(138)-ZSM-5 reacted at 500°C in 20% H$_2$O-2% NO-He for
                  10 hours, then in dry gas (2% NO-He) for 20 hours.

Hydrolyzed copper complexes might be formed when Cu-ZSM-5 is exposed to water vapor at reaction temperatures. These copper complexes are more mobile than Cu^{2+} cations alone due to lower surface charge density. Therefore, some of the copper complexes may migrate out of the zeolite cages, subsequently leading to CuO particle formation and catalyst deactivation, and Brönsted acid site formation on the sites left by the copper. Another possibility is zeolite dealumination leading to loss of active sites inside the zeolite cages and easier copper migration out of the channels. Using ^{27}Al-NMR, Grinsted, et al (*14*) observed significant dealumination of a Cu-ZSM-5 with Si/Al = 40 after the sample had been exposed to a 10% H2O- containing gas at 410°C for 113 hours. In the present study, the extent of dealumination has not yet been examined for the steamed Cu-ZSM-5 and M/Cu-ZSM-5 catalysts. However, little dealumination of the 500°C-steamed parent ZSM-5 (for 20 hours) seems to have occurred, based on the data in Table 2.

Upon water vapor removal from the stream, only part of the dry gas-catalytic activity is recovered. This partial activity recovery may be mainly attributed to decomposition of the hydrolyzed copper complexes, and bare ion (due to the dehydration of copper complexes on the exterior surface) migration to active sites inside the zeolite cavities. Another contribution may come from the fine CuO particles on the zeolite surface, i.e., small part of active Cu^{2+} cations are slowly restored by solid ion exchange with Brönsted acid sites. This hypothesis is drawn from the observation that Cu(H)-ZSM-5 with low Cu ion exchange level can be obtained by solid ion exchange between H-ZSM-5 and CuO in a vacuum at 500°C as reported by Karge, et al (*15*). The permanent activity loss, however, is not explained. This may be attributed to irreversible CuO particle formation and deactivation, or dealumination if it happened in this study.

The presence of the Ce cations in the catalyst Ce(60)/Cu(138)-ZSM-5 greatly improves the Cu-ZSM-5 wet gas-activity. Ce cations may stabilize the active copper cations against migration either directly or by stabilizing the framework aluminum. Any hydrated or hydrolyzed cerium or cerium particles on the external surface, may also keep the copper dispersed, thus, easier to migrate back, upon removal of the water vapor (Figure 8). This would decrease CuO particle sintering. Detailed mechanistic studies of the Ce cation promotion of the Cu-ZSM-5 activity and catalyst composition optimization are warranted in view of the results of the present study.

Conclusion

Experimental results show that Mg^{2+}, Ce^{3+} -modified Cu-ZSM-5 catalysts are more active than unpromoted Cu/ZSM-5 for the NO decomposition to nitrogen and oxygen, both in O2-free or O2-containing gas mixtures. The cocations are able to stabilize Cu^{2+} cations on active sites. Severe steaming (20% H_2O, 20 hours) at high temperatures ($\geq$ 600°C) causes zeolite dealumination and vitreous glass formation leading to loss of Cu cation uptake capacity and micropore volume loss. Wet NO decomposition at 500°C results in greatly reduced Cu-ZSM-5 activity due to the Cu migration and CuO particle formation. The presence of cocations moderates the effect of water vapor on the copper ion activity. Ce cations have been shown in this work to improve not only the activity recovery upon water vapor removal, but most importantly, the wet gas NO decomposition activity. Less extensive CuO particle formation on the surface of steamed Ce/Cu-ZSM-5 catalysts has been identified by STEM analysis.

Acknowledgments

The discussion of the XRD results with Prof. Bernhardt J. Wuensch of the MIT's Department of Materials Science and Engineering is greatly appreciated by one of us (Y. Zhang). This project was financially supported by the U.S. Department of Energy/University Coal Research Program, under Grant No. DE-FG22-91PC91923, the Gas Research Institute under contract No. 5093-260-2580 and the ENEL- Italian Power Authority, Milan, Italy.

Literature Cited

1. Iwamoto, M.; Furukawa, H.; Mine, Y.; Uemura, F.; Mikuriya, S.;Kagawa, S. *J. Chem. Soc. Chem. Commun.* **1986**, *15*, 1272 -73.
2. Iwamoto, M. in *Future Opportunities in Catalytic and SeparationTechnology* Misono, M. et al Ed.; Elsevier: Amsterdam, **1990**; 121-143.
3. Li, Y.; Hall, W. *J. Catal.*, **1990,** *129*, 202-215.
4. Li, Y.; Armor, J. *Appl. Catal.*, **1991,** *76*, L1.
5. Shelef, M. *Catal. Lett.*, **1992**, *15*, 305-310.
6. Kagawa, S.; Ogawa, H.; Furukawa, H.; Teraoka, Y. *Chem. Lett.*, **1991**, 407-410.
7. Zhang, Y.; Flytzani-Stephanopoulos, M., in *Environmental Catalysis* ; Armor, J., Ed.; ACS Symposium Series 552; ACS, Washington, DC, **1994**; 7-21.
8. Iwamoto, M.; Yohiro, H.; Tanda, K.; Mizuno, N.; Mine, Y; Kagawa, S., *J. Phys. Chem.*, **1991**, *95*, 3727.
9. Li, Y.; Hall, W. K., *J. Phys. Chem.*, **1990**, *94*, 6148.
10. Kharas, K. C. C.; Robota, H. J; Datye, A., in *Environmental Catalysis* ; Armor, J., Ed.; ACS Symposium Series 552; ACS, Washington, DC, **1994**; 39- 52.
11. Olson, D. H.; Kokotailo, Lawton, S. L., *J. Phys. Chem.* **1981**, *85*, 2238. Also, Aronson, M. T.; Gorte, R. J.; Farneth, W. E., *J. Catal.*, **1986**,*98*, 434.
12. Flytzani-Stephanopoulos, M.; Sarofim, A. F.; Zhang, Y., Quarterly Technical Progress Report No. 7 to DOE, **1993**, Grant No. DE-FG22-91PC91923.
13. Qin, L. C., *Ph.D. Thesis*, Department of Materials Science and Engineering, Massachusetts Institute of Technology, April 1994.
14. Grinsted, R. A.; Jen, H.-W.; Montreuil, C. N.; Rokosz, M. J.; Shelef, M., *Zeolites*, **1993**, *13*, 602.
15. Karge, H. G.; Wichterlová, B.; Beyer, H. K., *J. Chem. Soc. Faraday Trans.*, **1992**, *88*, 1345.

RECEIVED November 28, 1994

Chapter 12

X-ray Absorption Spectroscopic Study of Cu in Cu-ZSM-5 During NO Catalytic Decomposition

Di-Jia Liu[1] and Heinz J. Robota[2]

[1]AlliedSignal Research and Technology, 50 East Algonquin Road, Des Plaines, IL 60017–5016
[2]AlliedSignal Environmental Catalysts Company, P.O. Box 580970, Tulsa, OK 74158–0970

We studied the Cu oxidation state and coordination changes in Cu-ZSM-5 during the NO catalytic decomposition reaction using X-ray absorption spectroscopic techniques, XANES and EXAFS. An *in situ* reactor system was designed with which we can measure the relative NO decomposition rate while recording X-ray absorption spectrum. We observed that the Cu(I) $1s \rightarrow 4p$ transition intensity increases from zero to near saturation after Cu-ZSM-5 is activated in high purity He flow. Meanwhile, the number of oxygen atoms surrounding copper ion decreases from 4 to 2. Under direct NO decomposition condition, Cu(I) $1s \rightarrow 4p$ transition becomes weaker. Nevertheless, its integrated intensity is well correlated with the decomposition rate at several reaction temperatures. This finding supports the conjecture that Cu(I) participates in a redox mechanism during catalyzed NO decomposition in Cu-ZSM-5. The catalytic activities of overexchanged versus underexchanged Cu-ZSM-5 and the effect of oxygen in the reactant stream are also discussed.

Although thermodynamically unstable with respect to the diatomic elements, the decomposition of NO is an extremely slow process. While many catalytically active materials will facilitate the rupture of the N-O bond, the continuous process of catalysis is typically hampered by the retention of the adsorbed oxygen. Copper, when appropriately introduced into ZSM-5 zeolite, has been shown effective at catalyzing NO decomposition to the elements(*1,7*). Perhaps more relevant to industrial application, Cu-ZSM-5 was also found to have high selectivity in removing NO with the presence of hydrocarbons in excess O_2(*1-6*). Elucidating the mechanisms of these catalytic processes represents a portentous opportunity to develop a more detailed understanding of the structural aspects of this catalyst which confers the observed activity(*7-14*). Furthermore, experiments have to be designed to demonstrate the proposed mechanism. For example, in the direct

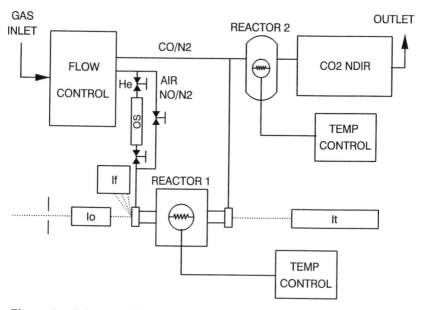

Figure 1.   Schematic diagram of the *in situ* reactor for X-ray absorption
spectroscopic study during NO decomposition over Cu-ZSM-5.

catalytic NO decomposition reaction, several mechanisms were proposed(*7-14*). Among them, Hall and co-workers suggested a redox reaction scheme in which the formation of Cu(I) by reductive desorption of oxygen from Cu(II) is a key step(*7,9*). A similar mechanism was also proposed by Iwamoto and Hamada(*2*). Several analytical techniques, such as FTIR, ESR, TPR, microbalance, etc., have been applied to address the question of the copper oxidation state and address its influence on the interaction of Cu with adsorbates under various conditions(*7-17*). In their microbalance experiment, for example, Li and Hall observed that the rate of spontaneous O_2 desorption by an oxidized Cu-ZSM-5 and the rate of NO decomposition at the same temperature, which were measured separately, are closely correlated. Based on this finding and other supporting evidence, they suggested that "the working catalysts probably consist of a relatively small number of Cu(I) centers (or adjacent Cu(I) pairs) maintained in a steady state by balance of the rate of oxidation with the rate of O_2 desorption"(*7*). To investigate this conjecture experimentally, it would certainly be very desirable to study simultaneously the distribution of copper oxidation states and rate of NO decomposition during actual reaction.

We have investigated recently the electronic and structural properties of Cu in Cu-ZSM-5 by the X-ray absorption near edge structure (XANES) and extended X-ray absorption fine structure (EXAFS) spectroscopic methods. Using the XANES and EXAFS spectroscopic techniques to study the oxidation state and local structure of transition metals is a well established method(*18*). Both techniques view the catalytic process directly from the perspective of the active site, Cu in ZSM-5 in this case, providing unique information which complements the results derived from surface spectroscopic studies or other indirect techniques. Since X-rays can penetrate the catalyst sample without disrupting the catalytic process, it is feasible to operate a catalyst at its reaction temperature and pressure while monitoring the electronic and local structure of the catalytic elements. We designed such a reactor system for NO decomposition study with which we can measure the relative NO conversion level while taking EXAFS and XANES spectra.

Experimental

The Cu-ZSM-5 samples in this study were made by exposing commercial H-ZSM-5 materials (typically with Si/Al atom ratios of 30) to aqueous cupric acetate solution. The details of the Cu exchange procedure is given in Ref. 19. This procedure can result in incorporation of cupric ions greater than the nominal exchange capacity of the zeolite. Throughout this text, we define a sample's name based on its exchanged level. For example, a Cu-ZSM-5-150 catalyst containing a cupric ion content of 150% of the nominal exchange capacity of the zeolite which, equivalently, contains a Cu/Al atom ratio of 0.75. The samples studied in this experiment are calcined in dry air at 673 K for two hours after ion exchange and filtration. No further aging process was involved.

The *in situ* reactor system for X-ray absorption spectroscopic study during NO decomposition is shown in **Fig. 1**. Reactor 1 is a flow-through type reactor where the sample pre-treatment and NO decomposition were carried out. During the experiment, Cu-ZSM-5 samples were packed as thin, permeable disks and placed in a sample holder at the center of the reactor which is surrounded by an insulated resistive heater. The sample holder was designed in such a way that all the reactant flow has to pass through the catalyst, similar to a plug-flow reactor. A well collimated X-ray beam passed through the X-ray intensity calibrator (I_0), sample holder, and entered the transmission detector (I_t) where the absorption of X-rays was recorded. At the entrance of Reactor 1 a small piece of Cu foil on a lead tape was placed to partially block the incident X-rays. The fluorescence excitation spectrum of the Cu foil was collected simultaneously by detector (I_f) during each scan to calibrate the Cu^0 edge energy. The zero of energy in this study is defined as the first inflection point of the copper metal foil fluorescence excitation spectrum. The temperature of Reactor 1 can be varied from room temperature up to 973 K. During the auto-reduction experiment, we used ultra-high purity He (Matheson, 99.9999%) treated with an on-line oxygen scrubber(OS) to eliminate any interference by residual oxygen or water. A 1% NO in N_2 mixture was used as the reactant gas during NO decomposition experiments. The flow rate, typically 100 ml/min, was regulated with a mass flow controller. The effluent stream, which contains N_2, O_2, NO_2, and unreacted NO, is combined with a stream of a 1% CO/N_2 mixture before entering Reactor 2 which contains a Cu,Sn/γ-Al$_2$O$_3$ catalyst. The temperature of Reactor 2 was controlled at 450 K throughout the experiment. Our laboratory tests showed that under these conditions, CO will preferentially react with the decomposition product O_2 and the secondary product NO_2 to form CO_2 over the Cu,Sn/γ-Al$_2$O$_3$ catalyst while NO is virtually unaffected. NO_2 is formed by partial reaction of product O_2 with some of unreacted NO in the effluent stream at low temperature, as suggested by Li and Hall[20]. To evaluate the influence of NO_2 on the conversion level measurement, we performed a laboratory test of CO oxidation in Reactor 2 by mixing known amounts of O_2 into the NO reactant stream. We observed that CO oxidation levels are stoichiometrically consistent with the added O_2 only, suggesting that NO_2 reacts with CO to form CO_2 and NO. Therefore, the amount of CO_2 measured corresponds directly with twice the amount of the O_2 formed during the decomposition. The concentration of CO_2 in the effluent is detected by a non-dispersive CO_2 infrared detector downstream of Reactor 2. Therefore, the relative NO conversion level can be monitored by the change of CO_2 content. The use of the secondary reactor and CO_2 NDIR instead of chemiluminescence NO$_x$ detection or GC provided us a more robust, portable, yet inexpensive approach. Since it is an indirect measurement, we limited our determination of NO conversion to relative values for comparison. A full scale description of the design and the performance of the *in situ* X-ray absorption spectroscopy reactor will be given in a separate publication[21].

The X-ray absorption experiments were performed at Beamline X-18B of the National Synchrotron Light Source (NSLS) at Brookhaven National Laboratory. The synchrotron ring current during these experiments was in the range of 110 to 210 mA and the ring energy was 2.53 GeV. A pair of Si(220) crystals was used in the monochromator and the slit width of the monochromator exit was 0.2 mm vertical and 15 mm horizontal to ensure optimal resolution. Since part of incoming X-ray beam was intercepted by copper foil for energy calibration, the actual horizontal beam width passing through the sample was 8 mm. All the X-ray absorption spectra were taken at the Cu K-edge (8979 eV) with near-edge energy resolution of 0.5 eV. For the EXAFS experiment, a typical spectrum covers a wave vector, k, range of 14 Å^{-1} for appropriate Fourier transformation.

Results and Discussion

Oxidation and Auto-Reduction of Cu in Cu-ZSM-5. In order to characterize the oxidation state and local geometric structure of copper ions during a redox reaction, we first studied these physical properties of Cu in fully oxidized and reduced Cu-ZSM-5. Shown in **Fig. 2(a)** is the XANES spectrum of Cu-ZSM-5-59. This sample has been exposed to air and moisture at room temperature for a period of time after it was prepared by ion exchange and calcination procedures. The up-shift of the edge energy and a very weak $1s \rightarrow 3d$ forbidden transition at near zero electron-volts indicate that the copper in ZSM-5 is in the form of cupric ion, Cu(II). The radial distribution function (RDF) of copper, obtained after the Fourier transformation of the EXAFS spectrum, is shown in **Fig. 3(a)**. The analysis shows that the cupric ion is surround by an average of 4.2 oxygen atoms in its first neighbor shell with shell radius of 1.96 Å, as is listed in Table I. No well-defined higher shell structure was observed. Relatively high 1st shell coordination number and lack of higher shell structure suggest that the Cu(II) ion in the zeolite is fully coordinated, possibly by O atoms in water and zeolite, and randomly distributed through the zeolite framework.

After calcination in dry air at 773 K for 30 minutes, the catalyst is fully dehydrated and oxidized. Earlier studies of Cu in zeolite Y suggested that cupric ions migrate to new sites upon dehydration and are coordinated by four oxygen atoms in a tetrahedral geometry(*22*). Although ZSM-5 is a very different zeolite, we also observed significant changes in both XANES and EXAFS spectra, as are shown in **Fig. 2(b)** and **Fig. 3(b)**. In the XANES region, the edge energy is slightly red-shifted and a weak transition appears at 8.5 eV above the zero which has been tentatively assigned as the $1s \rightarrow 4p$ transition of Cu(II). In addition, the $1s \rightarrow 3d$ forbidden transition of cupric ion maintains approximately its previous intensity. The RDF obtained from EXAFS analysis shows a well-defined second and third shell structure. The numerical fitting of the Fourier filtered first shell EXAFS indicates that the number of oxygen atoms in the 1st shell is 3.7, shown in Table I. All these observations converge to a single fact that copper ions in ZSM-5 remain as Cu(II) upon dehydration in dry air with a likely change in coordination related to anchoring of the ions to zeolite O^{-2} ions. The slight decrease in the 1st shell coordination number results from the loss of H$_2$O, which also accounts for the red shift of the edge energy because of higher electron affinity of water than that of the zeolite framework.

It has been demonstrated that, at elevated temperature, Cu(II) in ZSM-5 can be auto-reduced to Cu(I) in an inert gas flow or *in vacuo*. For example, Li and Hall showed the formation of Cu(I) from Cu(II) in He flow by ESR spectroscopy and demonstrated that the self-reduction results from the desorption of oxygen(*7*). By activating underexchanged Cu-ZSM-5 in ultra-high purity He flow at 773 K, we observed the growth of a narrow, intense peak at 5.3 eV above the Cu^0 K-edge

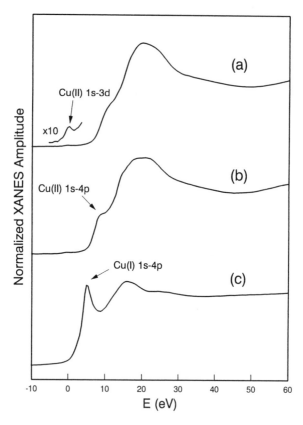

Figure 2. XANES spectra of an underexchanged Cu-ZSM-5-59 at the Cu K-edge: (a) sample has been exposed to ambient air after ion exchange and calcination; (b) sample was oxidized in dry air at 773 K and was cooled in dry air to room temperature; (c) sample was auto-reduced in ultra-high purity He at 773 K and was cooled to room temperature in He. The amplitude of each XANES spectrum is normalized based on the procedure provided in Ref. 18.

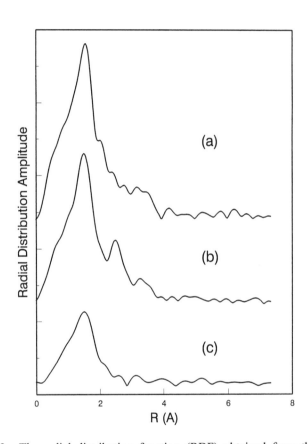

Figure 3. The radial distribution function (RDF) obtained from the Fourier transformation of EXAFS spectra for the underexchanged Cu-ZSM-5-59: (a) sample has been exposed to ambient air after ion exchange and calcination; (b) sample was oxidized in dry air at 773 K and was cooled in dry air to room temperature; (c) sample was auto-reduced in ultra-high purity He at 773 K and was cooled to room temperature in He.

in the XANES spectrum. The growth of the peak gradually stops after about one hour, indicating completion of auto-reduction from Cu(II) to Cu(I). **Fig. 2(c)** shows the XANES spectrum after the sample was cooled to room temperature in He flow. A similar sharp, intense feature has been previously observed in various copper compounds and is assigned to an electronic transition from $1s$ to $4p_{xy}$ in cuprous complexes(23). The relationship between the peak intensity and the coordination number in cuprous complexes has been systematically studied(23) and the peak intensity we observed in **Fig. 2(c)** suggests that Cu(I) is coordinated by only two ligands. The Fourier transformed EXAFS spectrum which is shown in **Fig. 3(c)** provides further evidence for a reduction in coordination. The numerical fitting of the 1st shell in the RDF reveals that the Cu(I) ion is surrounded by approximately two oxygen atoms, as is listed in Table I. This observation is consistent with the fact that Cu(I) in Cu-ZSM-5 is formed by desorption of oxygen (7) and the narrow $1s \rightarrow 4p$ transition is the result of two ligand coordination. According to the literature, the preferred geometry for 2-oxygen coordinated cuprous complexes has a dumbbell type shape, O-Cu-O(24).

Table I. The first shell coordination number N and shell radius R of Cu-ZSM-5-59 after the sample was treated on-line under different conditions.

	N	R (Å)	$\delta\sigma^2$ (Å^2)
Before calcination in air	4.5	1.95	0.0015
After calcination in air	3.6	1.93	0.0002
After auto-reduction in He	2.1	1.94	0.001

We should point out that Cu(I) in Cu-ZSM-5 formed through auto-reduction is extremely reactive with oxidizing species at room temperature. We observed during our experiments that although Cu(I) is formed at elevated temperature even with the presence of oxidizing species (NO, O$_2$ etc.) in the gas stream, the chemical balance shifts strongly towards Cu(II) when the reaction temperature approaches room temperature. For example, we observed a nearly 100% auto-reduction of copper ion in Cu-ZSM-5-59 at 773 K in a high purity N$_2$ (99.99%, Matheson) flow, the Cu(I) ions were later completely re-oxidized to Cu(II) during cooling by the trace amount of oxygen contaminant in the nitrogen. Therefore, to study the structure of Cu(I) in Cu-ZSM-5 at ambient temperature, it is crucial to remove trace amounts of oxidizing species in the gas flow. Furthermore, this observation underscores the importance of *in situ* study for Cu-ZSM-5 during NO decomposition and selective reduction by hydrocarbons. Because in both cases there exists a significant amount of oxidizing species in the reactant gas stream, it is impossible to treat the sample under reaction conditions and then study it by returning to room temperature without distorting the actual oxidation state and structural information of Cu in the zeolite.

Direct NO Catalytic Decomposition.   By studying the fully oxidized and auto-reduced copper ions in Cu-ZSM-5, we established a set of standard XANES spectra.   We can therefore analyze the content of cuprous and cupric ions by simulating the observed XANES spectrum with the superposition of fractional amounts of the standard Cu-ZSM-5 spectra.   For example, the solid line in **Fig. 4** is the XANES spectrum of an overexchanged Cu-ZSM-5-164 after activation in He at 773 K for 30 minutes.   The dashed line represents the linear summation of 70% of the XANES spectrum of Cu(I)-ZSM-5 and 30% of the XANES spectrum of Cu(II)-ZSM-5, both obtained at 773 K.   Excellent agreement between the experimental data and the simulation is observed throughout the entire XANES region (from threshold up to 100 eV above the edge) by this method.   The error of the fitting is ±5%.   This indicates that about 70±5% of the cupric ions in Cu-ZSM-5-164 were auto-reduced to Cu(I) and the rest remained as Cu(II).

The Cu(I) $1s{\rightarrow}4p$ transition intensity decreased drastically after the admission of the 1% NO/N_2 gas mixture into Reactor 1, indicating that a substantial amount of Cu(I) formed in the activation process was quickly oxidized back to Cu(II), possibly by O_2 formed during the decomposition reaction.   Although the Cu(I) transition intensity is weak at the reaction temperature, it is by no means completely extinguished.   Even after a prolonged period of catalytic reaction while reaching a steady state, there still exists a constant cuprous ion concentration at 773 K.   Using the spectral simulation analysis mentioned above, we calculated a Cu(I) content of approximately 35±5% of the total copper ion content.   Apparently, at elevated temperatures, a certain fraction of copper was maintained as cuprous ion even in the oxidizing environment.

Correlation between Cu(I) Concentration and NO Decomposition Rate.
To explore the relationship between the Cu(I) content and the decomposition activity proposed by Li and Hall(7), we conducted the reaction in a temperature cycle around the optimum conversion temperature of 773 K and recorded the XANES spectra *in situ* at each temperature.   The temperature cycle used was: T_1 = 773 K $\rightarrow$ T_2 = 873 K $\rightarrow$ T_3 = 773 K $\rightarrow$ T_4 = 673 K $\rightarrow$ T_5 = 573 K $\rightarrow$ T_6 = 773 K with an approximate space velocity (GHSV) of 80,000 hr^{-1} at 773 K.   At this space velocity, NO decomposition was carried out in the differential regime and was not diffusion limited.   The NO conversion level varied with temperature and reached a steady state after 30 minutes operation at a particular temperature.   A maximum conversion level at 773 K was observed, as was previously reported(1).   Correspondingly, we also found that the Cu(I) $1s{\rightarrow}4p$ peak intensity changed throughout the temperature cycle.   Shown in **Fig. 5** are the normalized XANES spectra taken at each temperature.   We observed that the intensity of the $1s{\rightarrow}4p$ transition, which is proportional to the population of Cu(I) in Cu-ZSM-5, indeed depends rather strongly on the reaction temperature.   We calculated the integral intensity of these peaks by using a normalized XANES difference method(23) and plotted them in **Fig. 6** on a relative scale with the normalized NO decomposition rate which we calculated from the relative conversion level at each point of the reaction cycle.   In order to study the correlation between the Cu(I) concentration and the relative NO decomposition rate, we used only the relative Cu(I) concentration here.   To obtain the absolute cuprous ion content one can simply

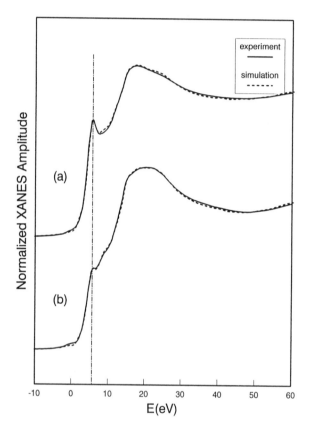

Figure 4. The normalized XANES spectra of Cu-ZSM-5-164 at T=773 K. The sample is pretreated in He flow first (a) and then are exposed to 1%NO/N$_2$ mixture (b). The solid lines represent the experimental spectra and the dashed lines represent the simulation by linear summation of XANES spectra of fully reduced and fully oxidized Cu in Cu-ZSM-5.

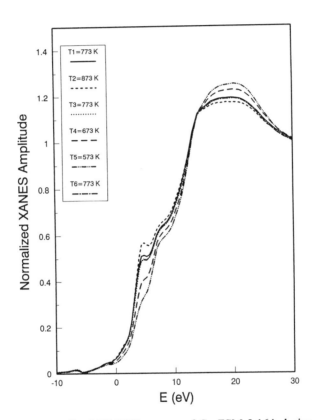

Figure 5. The normalized XANES spectra of Cu-ZSM-5-164 during the NO decomposition temperature cycle.

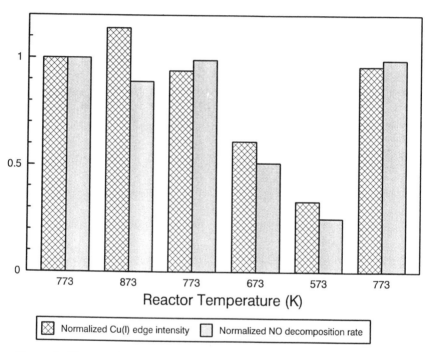

Figure 6. The correlation between the Cu(I) pre-edge transition intensity with the relative NO decomposition rate in the temperature cycle. Both intensity and the decomposition rate are normalized to their respective values at T_1.

scale the relative value with that obtained at 773 K, which has 35% Cu(I) under NO decomposition condition. For the method to calculate the normalized difference edge absorption spectrum, see Ref. 23.

The good correlation between the Cu(I) content and NO decomposition rate is noteworthy. Since the measurement was carried out under reaction conditions, it confirms the conjecture by Hall *et al.* that a small fraction of cuprous ion is maintained in the working catalyst(7, 9). Both the cuprous ion concentration and the catalyst efficiency vary in the same way with temperature. According to the redox mechanism proposed by Li and Hall(7), the rate of NO decomposition should be proportional to the partial pressure of NO and the number of the available active sites in Cu-ZSM-5, which they attribute to Cu(I) ions. The correlation that we observe seems to support their notion that Cu(I) participates in the mechanism of NO decomposition at elevated temperature.

It is interesting at this point to compare our experimental results with those of Li and Hall(7). They proposed that the following two reactions form the redox cycle of the NO decomposition over Cu-ZSM-5;

$$NO + (S) \rightarrow NO\text{-}(S) \rightarrow \frac{1}{2}N_2 + O\text{-}(S) \tag{1}$$

and

$$2O\text{-}(S) \rightleftharpoons O_2 + 2(S), \tag{2}$$

where (S) represents surface vacancies in reduced form, i. e. Cu(I); NO-(S) and O-(S) represent adsorbed NO and oxygen atom sites, respectively. As part of their experimental approach to the redox mechanism, they measured the weight loss of oxidized Cu-ZSM-5-166 in He flow from 623 K to 823 K and attributed the loss to the desorption of dioxygen. In a separate experiment, they also measured the turnover frequency of NO decomposition at the same temperatures. They observed that the rates of these two distinct processes were correlated, (Table 4 of Ref. 7). Since the desorption of O_2 will lead to the reduction of Cu(II) to Cu(I) which is believed to be the active site (S), they suggest that the rate of cuprous ion formation is associated with the rate of NO decomposition and the oxygen desorption process in Eqn. 2 is the rate limiting step in the redox cycle.

The oxygen desorption rate provides information on the forward reaction of Eqn. 2; i. e., how fast the cuprous ion in Cu-ZSM-5 regenerates by auto-reduction. On the other hand, the Cu(I) in the catalyst during reaction will be continuously depleted by the oxidizing agents, as is shown in Eqn. 1 and the backward reaction of Eqn. 2. Therefore the cuprous ion content is actually controlled by a dynamic balance of the desorption of oxygen, adsorption of NO as well as the re-oxidation by product oxygen. The rate constants in Eqn. 1 and 2, and therefore the steady state concentration of Cu(I), should depend on the reaction temperature. The presence of a significant amount of Cu(I) during NO decomposition, e.g. 35% of all copper species at 773 K, indicates that although the forward reaction in Eqn. 2

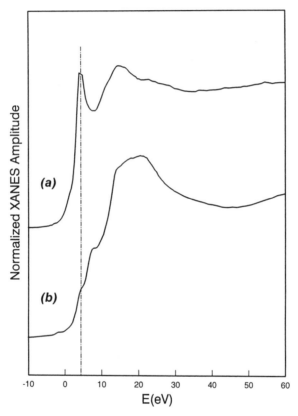

Figure 7. The normalized XANES spectra of Cu-ZSM-5-59 at T=773 K. The sample is pretreated in He flow first (a) and then is exposed to the 1%NO/N$_2$ mixture (b).

may be the rate limiting step in the redox cycle, it is not much slower than the re-oxidation and decomposition reactions. Otherwise we would see much less steady state Cu(I) due to fast depletion by reaction in Eqn. 1 and backward reaction in Eqn. 2. In fact, a reverse process in Eqn. 1, i.e. NO desorption from NO-(*S*), should be also included. The overall rate constant for direct NO decomposition contains probably more reaction processes than were included in Ref. 7. In **Fig. 6** we observe that the correlation between cuprous ion concentration and NO decomposition rate breaks down at 873 K; the Cu(I) intensity reaches a maximum yet the decomposition rate decreases slightly from that at 773 K. Similar behavior was also observed in Li and Hall's study where they reported a higher oxygen desorption rate at 827 K than at 773 K, although the decomposition rate decreased. These observations suggest that the rate of Cu(I) regeneration ceases to be the rate determining step at higher temperature. Otherwise the decomposition rate would continue to increase with temperature. The NO adsorption, which leads to the formation of a reactive intermediate with Cu(I), may decline in rate at very high temperature, possibly due to fast desorption. It has been observed in FTIR experiments that the intensities of several infrared bands assigned to NO adsorbed on Cu(I) sites decrease rapidly with increasing temperature(*25*). The mononitrosyl $Cu^{(1+\delta)+}(NO)^{\delta-}$ and dinitrosyl $Cu^{(1+\delta)+}(NO)_2^{\delta-}$ species are suggested as the reaction intermediates for NO decomposition. Similar to oxygen adsorption, the formation of nitrosyl intermediates will oxidize Cu(I) to cupric ion through electron transfer. The fact that we detected more Cu(I) at 873 K suggests that the rate at which reactive intermediates are formed is outpaced by the NO desorption and auto-reduction rates. The negative temperature dependance of the NO decomposition rate at T > 773 K may simply indicate that Eqn. 1 becomes the rate-limiting step and its rate constant declines rapidly with temperature.

Overexchanged versus Underexchanged Cu-ZSM-5. In their study on direct NO decomposition, Iwamoto *et al.* found that the catalytic activities of Cu-ZSM-5 increase with copper ion exchange level and level off over 100% exchange capacity(*1*). This observation was confirmed by Li and Hall in their kinetics study(*7*). They also observed that oxygen desorption rate for Cu-ZSM-5-166 is higher than that of Cu-ZSM-5-76. From the exchange site distance analysis and UV spectroscopic study, Iwamoto *et al.* proposed that a dimeric copper species, possibly in the form of $Cu^{2+}-O^{2-}-Cu^{2+}$, is the precursor for auto-reduction of the complete redox cycle.

We studied the XANES spectra of an over exchanged Cu-ZSM-5-164 and an underexchanged Cu-ZSM-5-59. Both catalysts were treated in He flow at 773 K, 30 minutes for 164% and 45 minutes for 59% exchanged samples, before He was replaced by 1% NO/N$_2$. Their corresponding XANES spectra are shown in **Fig. 4** and **Fig 7**. The most profound feature in these spectra is the difference in Cu(I) $1s \rightarrow 4p$ transition intensity before and after the admission of NO. For Cu-ZSM-5-164, about 70% of the copper ions were reduced to cuprous ions in He flow, using the aforementioned spectral analysis method. The Cu(I) percentage concentration decreases rapidly to 35% after the gas switch and maintains at this level during 60 minutes of NO decomposition catalysis at 773 K. For Cu-ZSM-5-59, virtually all the copper ions were auto-reduced to Cu(I) in He flow. The static

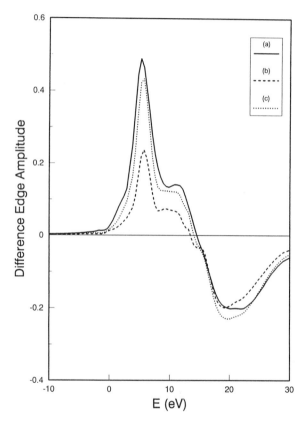

Figure 8. The normalized difference edge XANES spectra of Cu-ZSM-5-164 during a NO decomposition reaction cycle at T=773 K in (a) 1%NO/N$_2$, (b) 1%NO/N$_2$+4%O$_2$ and (c) back to 1%NO/N$_2$.

concentration of cuprous ions, however, drops to about 10% after He is switched to 1% NO/N_2.
The difference in Cu(I) contents between 164% and 59% exchanged zeolite samples during NO decomposition catalysis is significant. It indicates that overexchanged Cu-ZSM-5 tends to sustain higher Cu(I) concentration therefore higher catalytic activity. It is also consistent with the hypothesis of a redox mechanism. The question is, how is this accomplished during the NO decomposition process? The observations of higher O_2 desorption rate(7) and possibly ionic copper dimer(1) suggest that cuprous ion may be regenerated by a dicopper process. At elevated temperature, a certain fraction of the cupric moieties migrate on the surface of the zeolite channel. A close encounter of two cupric moieties may form a transient dicopper species and cuprous ions can be thus generated through the desorption of oxygen. The overexchanged Cu-ZSM-5 has higher surface Cu density, therefore can regenerate cuprous ion faster. The migration of transition metal ions in zeolite is a well known process(22). It eventually leads to the agglomeration of ionic copper to form copper oxides. To confirm this speculation, however, more experimental studies, especially time-resolved X-ray absorption spectroscopic study, should be conducted.

So far, we speculated that the catalytic activity difference between the overexchanged and underexchanged Cu-ZSM-5 during NO decomposition may be due to the efficiency of Cu(I) regeneration. This mechanism, however, can be altered rather significantly if the reaction environment changed. For example, the underexchanged Cu-ZSM-5 is a rather active catalyst in the selective NO reduction by hydrocarbon. In view of the fact that hydrocarbon is a very effective reducing agent, a different mechanism arises which is presented in a separated publication(26).

Effect of O_2 on NO Decomposition. Addition of oxygen to the gas stream will partially suppress the NO decomposition activity on Cu-ZSM-5. According to Iwamoto *et. al.*(1), the degree of suppression depends on the partial pressure of O_2 and it can be rather significant with the presence of high oxygen content in the stream. It is also observed that the catalyst returns to its original activity after the oxygen is removed. We studied the XANES of Cu-ZSM-5-164 by adding 4%O_2 in 1%NO/N_2 mixture in our *in situ* reactor. Due to the intrinsic limitation of the reactor design, we could not monitor the NO conversion level in this case. Instead, only XANES spectra were recorded. We plot the normalized difference edge XANES spectra, as are shown in **Fig. 8**, in order to emphasize the small variation of the Cu(I) intensity during the reaction. We observed that adding O_2 in the flow reduced the Cu(I) content substantially in Cu-ZSM-5, as is shown in **Fig. 8a** and **b**. However, there is still a small fraction of Cu(I) ion in the zeolite under this condition. After removing O_2, the Cu(I) transition was gradually restored to approximatedly its original level, as is shown in **Fig. 8c**. This experimental fact is consistent with the conjecture that Cu(I) participates in the overall reaction scheme. The suppression of decomposition activity is associated with the depletion of Cu(I) concentration.

Acknowledgements

We thank Dr. Karl C. C. Kharas for providing the samples and Mr. Mike A. Reddig for technical assistance during this investigation. Financial support of AlliedSignal Research and Technology is gratefully acknowledged. Research carried out (in part) at X18B of National Synchrotron Light Source, Brookhaven National Laboratory, which is supported by the U. S. Department of Energy, Division of Materials Sciences and Division of Chemical Sciences.

Literature Cited

1. For example, Iwamoto, M.; Mizuno, N.; Yahiro, H. *Sekiyu Gakkaishi*, **1991**, *34*, 375, and references therein.
2. Iwamoto, M.; Hamada, H.; *Catal. Today*, **1991**, *10*, 57.
3. Inui, T.; Kojo, S.; Shibata, M.; Yoshida, T.; Iwamoto, S. *Stud. Surf. Sci. Catal.*, **1991**, *69*, 355.
4. Teraoka, Y.; Ogawa, H.; Furukawa, H.; Kagawa, S. *Catal. Lett.* **1992**, *12*, 361.
5. Iwamoto, M.; Yahiro, H.; Tanda, K.; Mozino, N.; Mine, Y.; Kagawa, S. *J. Phys. Chem.* **1991**, *95*, 3727.
6. Held, W.; König, A.; Richter, T.; Puppe, L. *Society of Automotive Engineers*, paper 900496.
7. Li, Y.; Hall, W. K. *J. Catal.* **1991**, *129*, 202.
8. Iwamoto, M.; Yahiro, H.; Mizuno, N.; Zhang, W.-X.; Mine, Y.; Furukawa, H.; Kagawa, S. *J. Phys. Chem.* **1992**, *96*, 9360.
9. Hall, W. K.; Valyon, J. *Catal. Lett.* **1992**, *15*, 311.
10. Kucherov, A. V.; Gerlock, J. L.; Jen, H.-W.; Shelef, M. *J. Phys. Chem.* **1994**, *98*, 4892.
11. Giamello, E.; Murphy, D.; Magnacca, G.; Morterra, C.; Shioya, Y.; Nomura, T.; Anpo, M. *J. Catal.* **1992**, *136*, 510.
12. Spoto, G.; Bordiga, S.; Scarano, D.; Zecchina, A. *Catal. Lett.* **1992**, *13*, 39.
13. Valyon, J.; Hall, W. K. *J. Phys. Chem.* **1993**, *97*, 1204.
14. Bell, A. T. *Prep. Am. Chem. Soc. Div. Petro. Chem.* **1994**, *39(1)*, 98.
15. Li, Y.; Armor, J. N.; *Appl. Catal.* **1991**, *76*, L1.
16. Sárkány, J.; d'Itri, J. L.; Sachtler, W. M. H.; *Catal. Lett.* **1992**, *16*, 241.
17. Kucherov, A. V.; Slinkin, A. A.; Kondrat'ev, D. A.; Bondarenko, T. N.; Rubinstein, A. M.; Minachev, Kh. M.; *Zeolite*, **1985**, *5*, 320.
18. See, for example, *X-ray Absorption, Principles, Applications, Techniques of EXAFS, SEXAFS and XANES*, Koningsberger, D. C. and Prins, R. eds. John Wiley and Sons, New York, 1988.
19. Kharas, K. C. C.; *Appl. Catal B:* **1993**, *2*, 207.
20. Li, Y.; Hall, W. K. *J. Phys. Chem.* **1990**, *94*, 6145.
21. Liu, D.-J.; Robota, H. J. *Catal. Lett.* **1993**, *21*, 291. Liu, D.-J.; Robota, H. J. *to be published*.
22. Coudurier, G.; Decamp, T.; Praliaud, H. *J. Chem. Soc. Faraday Trans.* **1982**, *78*, 2661.

23.   Kau, L-S.; Spira-Solomon, D. J.; Penner-Hahn, J. E.; Hodgson, K. O.; Solomon, E. I. *J. Am. Chem. Soc.* **1987,** *109*, 6433.
24.   Müller-Buschbaum, H. *Angew. Chem. Int. Ed. Engl.* **1991,** *30*, 723.
25.   Valyon, J.; Hall, W. K. *Proc. 10th int. Congr. Catal.* (Budapest, 1992).
26.   Liu, D.-J.; Robota, H. J. *Appl. Catal. B:* **1994,** *4*, 155.

RECEIVED November 22, 1994

Chapter 13

X-ray Photoelectron Spectroscopy
of Metal-Exchanged Na-ZSM-5 Zeolites

L. P. Haack, C. P. Hubbard, and M. Shelef

Ford Motor Company, Mail Stop 3061, SRL, P.O. Box 2053,
Dearborn, MI 48121−2053

Medium-pore ZSM-5 zeolites exchanged with divalent and trivalent transition metal ions have been considered recently by many investigators as catalysts for the selective reduction of nitrogen oxides. In this work a series of NaZSM-5 samples exchanged by divalent ions Co^{2+}, Ni^{2+}, Pd^{2+}, Cu^{2+}, and trivalent ions Fe^{3+}, La^{3+} and Ce^{3+} was examined in situ by XPS. In the high Si/Al ratio ZSM-5 the divalent ions exchanged approximately one to one with respect to the Al^{3+}-ions. The exchanged cobalt ions remained in the zeolite lattice even after high temperature (600°C) alternating oxyreductive treatments with oxygen and hydrogen. Copper migrates out of the framework during reductive treatment with hydrogen and is reduced to the metallic state. Nickel and palladium appear to migrate out to the external surface of the zeolite crystallites during high temperature oxidation and, subsequently, are reduced completely by treatment in hydrogen. Iron exchanged from a ferric solution is incorporated as ferric ions after oxidative treatment and is reduced to the ferrous state by hydrogen at high temperature without migrating out of the framework. The trivalent La-ions remain invariant in all treatments. The Ce-ions are present mainly as tetravalent after the oxidative treatment and are reduced completely to trivalent after treatment in hydrogen at 600°C. Both La- and Ce-ions appear to resist migration out of the framework. No difference could be detected in the behavior of Cu-ions exchanged into HZSM-5, NaZSM-5 and LaNaZSM-5.

The exchange of metallic ions into zeolites, and into ZSM-5 in particular, is a subject of considerable interest in practical catalysis since there are at present several examples of ZSM-5 catalysts having specific properties where the protons

0097−6156/95/0587−0166$12.00/0
© 1995 American Chemical Society

have been exchanged for other positive ions. Among these are GaHZSM-5 and ZnHZSM-5 in aromatization of lower alkanes (*1,2*), NiHZSM-5 in the production of paraxylene by isomerization of C_8-alkyl aromatics (*3*), CoHZSM-5 and FeHZSM-5 in the indirect liquefaction of synthesis gas (*4*), and CuHZSM-5 and several other metallo-zeolites in the selective reduction of NO_x (*5,6*). The exchange of trivalent ions such as La^{3+}, Ga^{3+}, In^{3+} and Ce^{3+} has been claimed as well (*7,8*). The introduction of some cations in the ZSM-5 is known to stabilize the material against dealumination by steam at high temperatures (*9,10*) which is important for the implementation of zeolite-based SCR catalysts into practice (*11*).

The exchange process of multivalent ions in high Si/Al ratio ZSM-5, where the protons are well isolated from each other, requires neutralization of the multiple charges by out-of-lattice negative charges. The higher the charge of the incoming ion, the lower is the probability of proton exchange. We sought to characterize the state of various ions incorporated into NaZSM-5 by surface analysis methods. This paper summarizes the XPS results. We will report subsequently on the characterization of the same materials by magic angle spinning nuclear magnetic resonance (MAS NMR) spectroscopy and on the catalytic behavior in NO_x SCR of some of the samples.

Experimental

Sample preparation. The starting material was NaZSM-5 from Alcoa. The choice of the starting material was dictated by the ease of exchange, considered to be higher with NaZSM-5 than with HZSM-5. The drawback is the presence of Na residue. The X-ray fluorescence (XRF) analysis of the NaZSM-5 gave the following result, on an oxygen- and water-free basis: Na-3.9, Al-3.9, Si-92 wt.-%. There is an excess of Na present, approximately 15%, external to the lattice. The calculated SiO_2/Al_2O_3 ratio is 46. The exchange, or addition, because in some cases exchange was not achieved, of the stabilizing ions was done from the nitrate solutions of the respective ions. The introduced amounts were aimed to be somewhat below (80%) the atomic content of the tetrahedral Al-ions. In a typical example, say La^{3+}-ions, 0.5 liter of a 0.007 molar solution of lanthanum nitrate in water was contacted at a pH of 5.5 with 9.5 g of NaZSM-5, heated with stirring at 85°C for 2 h, cooled to ambient, filtered and washed with water, dried in air at 140°C for 2 h and calcined in air for 2 h at 500°C. The solid was then recontacted with the mother liquor and the filtering, washing, drying and calcining were repeated. The added ions included Fe^{3+}, Co^{2+}, Ni^{2+}, Ce^{3+}, La^{3+}, Pd^{2+} and Cu^{2+}. One composite sample was also prepared by taking the finished LaNaZSM-5 and exchanging the residual Na^+ with copper nitrate solution in a similar manner as above, i.e. twice.

Table I summarizes the bulk composition of the exchanged samples determined after preparation. All metals, except Na, were determined by inductively coupled plasma atomic emission spectroscopy (ICP-AES). Na was determined by XRF analysis. Table I gives the residual Na content, the content of the introduced metal ions (on an oxygen- and water-free basis) and the atomic ratio of the introduced ions to Al^{3+}-ions. Of more significance is the last column which gives the atomic ratio of the introduced ions to the "free" Al^{3+}-ions, i.e. those not associated with residual Na^+-ions. This is an approximate figure because the original NaZSM-5 contained some excess Na and we do not have precise data how much of this excess was washed out of the sample in the preparation procedure. It is clearly seen that the atomic ratio of the four divalent ions to "free" Al^{3+} is about one (0.81-1.2). The same ratio for the 3 trivalent ions is 0.32 to 0.55.

Table I. Bulk composition of metal-exchanged NaZSM-5 samples[a]

Sample	Wt.-% Na Remaining	% Na Remaining	Wt.-% Metal Exchanged	At. Ratio Metal/Al	At. Ratio Metal/ (Al-Na)
FeNaZSM-5	0.77	36	1.5	0.19	0.32
CoNaZSM-5	1.6	41	4.0	0.48	0.93
NiNaZSM-5	1.8	46	4.2	0.51	1.1
CeNaZSM-5	1.7	44	5.2	0.27	0.55
LaNaZSM-5	1.8	46	4.8	0.25	0.54
PdNaZSM-5	0.60	15	8.3	0.58	1.2
CuNaZSM-5	0.84	22	5.4	0.61	0.81
CuLaNaZSM-5	0.66	17	3.0	0.35	0.43

[a]After initial calcination on an oxygen- and water-free basis.

If the ions were introduced initially by exchange then one Na^+-ion is being exchanged by one divalent ion, whose second charge is compensated by an extra-lattice charge. The trivalent ions can be accommodated only by some fraction of such sites. The composite sample had an atomic ratio of the sum of the introduced ions (La^{3+} + Cu^{2+}) to "free" Al^{3+} close to one as well. A sample of CuHZSM-5 was also examined by XPS; it is not included in Table I. The Cu content of this sample was 1.92 wt%.

XPS Analytical System. XPS Spectra were obtained using an M-Probe ESCA spectrometer manufactured by Surface Science Instruments, VG Fisons, with monochromatic $AlK\alpha$ X-rays (1486.7 eV, 80W) focused to a 1200-μm diameter beam. A low energy (1-3 eV) electron gun and a Ni charge neutralization screen placed 1-2 mm above the sample were employed to minimize surface charging effects (12). The analyzer was operated at a 50-eV pass energy. Binding energies were referenced to the zeolite Si 2p line at 102.9 eV, and measured to an accuracy of ± 0.3 eV. Metal/Si atomic ratios were determined by using photoionization yields of the metal core level and zeolite Si 2p line normalized by means of routines based on Scofield's photoionization cross-section values (13).

The zeolites were treated in a PHI Model 04-800 Reactor System, mounted directly onto the introduction chamber of the spectrometer, at 600°C and atmospheric pressure for 4 h. The reactor gases used were Ar (99.9995%), O_2 (99.98%) and H_2 (99.9995%), purchased from Matheson. A detailed description of the sample treatment has been given previously (*14*). Ce and Ni exchanged NaZSM-5 were additionally reacted with air (following XPS analysis after the reduction cycle) by removing the samples from the spectrometer and exposing them to the atmosphere for 1 min. A NiO reference sample was made by reacting Ni metal foil (0.025 mm thick, 99.99$^+$ purity, Goodfellow Metals) with a stream of O_2/Ar (50/50) flowing at 100 cm^3/min for 2 h at 600°C. A Ni metal reference sample was obtained similarly by reducing the NiO foil in a stream of H_2/Ar (50/50). After reactor treatments, all samples were transferred to the spectrometer *in vacuo* to eliminate contamination and oxidation by exposure to air.

Results and Discussion

All metal-exchanged zeolites were subjected to the sequence of reactions consisting of oxidation at 600°C with O_2, reduction at 600°C with H_2, exposure to air, and finally reoxidation at 600°C with O_2. For the iron, cobalt, lanthanum, palladium, and copper exchanged zeolites, all pertinent discriminating chemical and structural information was obtained after just the initial oxidation and reduction reactions. Therefore, XPS core level spectra for these zeolites are included only for the first two reactions. Additional information was revealed in the nickel and cerium exchanged zeolites after exposure to air, and then in the final high temperature reoxidation. Thus XPS spectra acquired after all chemical reactions are included for the nickel and cerium exchanged zeolites. We present XPS spectra for each sample acquired after various *in situ* treatments in Figures 1-6, and report the observed Metal/Si atomic ratios in Table II. This table also gives the same ratio for the bulk composition calculated from the ICP-AES analysis.

Iron-exchanged NaZSM-5. The Fe 2p core level spectra of the Fe-exchanged zeolite obtained after heating to 600°C in O_2, and then to 600°C in H_2, are shown in Figure 1. The spectrum acquired after oxidation was similar to that obtained by McIntyre and Zetaruk (*15*) for Fe_2O_3 oxide, which revealed a small satellite band about 8 eV above the main core line. However, the Fe $2p_{3/2}$ core level BE (711.2 eV) was about 0.4 eV higher for the iron-exchanged zeolite. Yue *et al.* (*16*) identified Fe^{3+} ions in a tetrahedral environment within the framework of a ZSM-5 zeolite synthesized by hydrothermal crystallization in an alkaline medium. They noticed a 0.9 eV higher BE shift for Fe^{3+} in the zeolite lattice over Fe^{3+} in fine Fe_2O_3 particles on the ZSM-5 surface.

After reductive treatment the Fe $2p_{3/2}$ BE was shifted down to 710.3 eV, and a distinct satellite band appeared 5.4 eV above the main core line. The shift to lower BE and prominent satellite structure are indicative of Fe^{2+} (*15*). If iron was present at the zeolite surface as non-interacting iron oxide particles, it would have been readily reduced to the metallic state (BE = 707 eV (*17*)) during the relatively harsh reductive treatment (*4*). Thus it is clear that all the iron probed by XPS exists interstitially within the pores of the zeolite structure and is, most probably, exchanged. Even so, it can be seen in Table II that the observed Fe/Si ratio by XPS, 3.1×10^{-2} after reduction, was considerably higher than that

Table II. Metal/Si atomic ratio of exchanged NaZSM-5

Sample	Treatment	Metal/Si Ratio × 10²	
		XPS	ICP-AES[a]
FeNaZSM-5	600°C O$_2$	3.8	0.83
	600°C H$_2$	3.1	
CoNaZSM-5	600°C O$_2$	2.1	2.1
	600°C H$_2$	2.0	
NiNaZSM-5	600°C O$_2$	13	2.2
	600°C H$_2$	3.2	
	R.T. Air	3.0	
	600°C O$_2$	5.2	
CeNaZSM-5	600°C O$_2$	1.6	1.2
	600°C H$_2$	2.2	
	R.T. Air	2.4	
	600°C O$_2$	2.1	
LaNaZSM-5	600°C O$_2$	1.8	1.1
	600°C H$_2$	1.6	
PdNaZSM-5	600°C O$_2$	9.1	2.5
	600°C H$_2$	7.3	
CuNaZSM-5	600°C O$_2$	2.5	2.7
	600°C H$_2$	3.7	
CuLaNaZSM-5	600°C O$_2$	0.59[b]	1.5
	600°C H$_2$	2.6[b]	

[a]After initial calcination.
[b]Cu only.

measured for the bulk (0.83×10^{-2}), suggesting a higher degree of exchange at the outermost layers of the zeolite.

Cobalt-exchanged NaZSM-5. The XPS spectra from the cobalt-exchanged zeolite are shown in Figure 2. The Co $2p_{3/2}$ BE for the oxidized sample was at 781.8 eV, while the spin-orbit split of the $2p_{3/2}$ and $2p_{1/2}$ peaks was 15.8 eV. A satellite peak was present about 5 eV above the main core line. In an XPS study of cobalt oxides Chuang *et al.* (*18*) reported a spin-orbit splitting for the Co $2p_{3/2}$ and $2p_{1/2}$ lines as 15.9 eV for CoO and 15.0 eV for Co_3O_4. In addition McIntyre and Cook (*19*) have demonstrated that a satellite is present in Co^{2+}, and not in Co^{3+}. The oxidation state of cobalt in the oxidized zeolite (Figure 1A) is clearly +2. The Co $2p_{3/2}$ BE at 781.8 eV is identical to that reported by Chin and Hercules (*20*) for dispersed cobalt on γ-Al_2O_3 after calcination. Chin and Hercules noticed that the BE measured for the dispersed cobalt was about 1.2 eV higher than that for Co_3O_4, and attributed the higher BE to a surface species closely resembling $CoAl_2O_4$. Also, Wei and Ying (*21*) measured a BE of 782.0 eV for cobalt-impregnated ZSM-5, and concluded that a $CoAl_2O_4$ species was contained in the interior of the zeolite.

It should be noted that BE's as high as 783 eV have been reported by others (*22,23*) for cobalt occupying ZSM-5 framework sites. However, in those studies the BE shift noticed between surface and framework cobalt was about 1 eV, the same as that observed by Wei and Ying (*21*). The differences in absolute BE's measured between the different authors can be attributed to discrepancies in referencing.

The spectrum in Figure 2B reveals that after reductive treatment only a slight amount (ca. 5%) of the cobalt was reduced to Co^0, as evidenced by the lower BE peak appearing at 778.2 eV (*24*). In a study of Co/ZSM-5 catalysts Stencel *et al.* (*22*) demonstrated that non-reducible Co^{2+} exists in the interior of the ZSM-5, while cobalt oxide particles on the exterior surface are reducible. The same was noted by Wei and Ying (*21*) and Rao (*4*) for cobalt-exchanged ZSM-5. Thus it is apparent that for the present study, most cobalt observed by XPS was exchanged into the zeolite structure. The small fraction of cobalt oxide reduced was most likely was present at exterior surface of the crystallites. The Co/Si atomic ratios by XPS and ICP-AES are identical (Table II), indicating a uniform exchange of the cobalt across the zeolite.

Nickel-exchanged NaZSM-5. XPS Ni 2p core level spectra acquired following oxidation, reduction, exposure to air, and then reoxidation of the nickel-exchanged zeolite are given in Figure 3, along with reference spectrum acquired from a NiO foil. The NiO standard exhibited a Ni $2p_{3/2}$ core line at 853.8 eV, with satellite lines at 855.3 and 860.9 eV. The nickel spectrum obtained from the oxidized zeolite was clearly different from that of the NiO standard. The zeolite spectrum showed a higher BE core line at 855.3 eV (same BE as the first satellite line of NiO) and a single satellite line at 860.9 eV, and matches well with the spectrum reported for Ni_2O_3 (*25*).

Figure 3B shows the Ni 2p spectrum acquired after reduction. The spectrum appears identical to that acquired from a Ni metal foil reference material (Ni $2p_{3/2}$

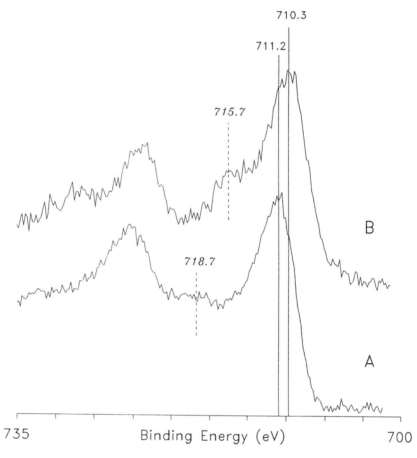

Figure 1. XPS Fe 2p core level spectra of FeNaZSM-5 A) after oxidation in 10% O$_2$/Ar at 600°C for 4 h and B) after reduction in 10% H$_2$/Ar at 600°C for 4 h.

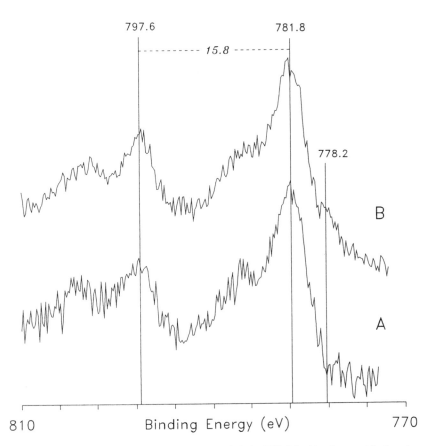

Figure 2. XPS Co 2p core level spectra of CoNaZSM-5 A) after oxidation in 10% O_2/Ar at 600°C for 4 h and B) after reduction in 10% H_2/Ar at 600°C for 4 h.

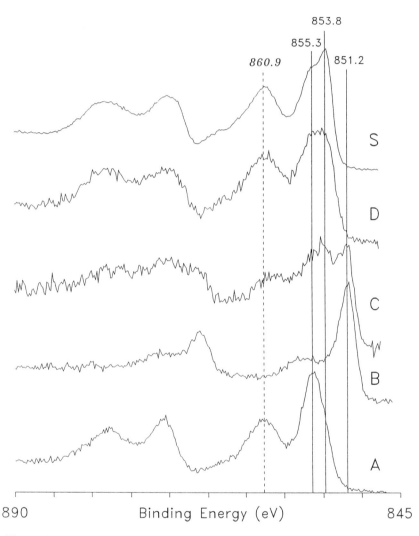

Figure 3. XPS Ni 2p core level spectra of NiNaZSM-5 A) after oxidation in 10% O_2/Ar at 600°C for 4 h, B) after reduction in 10% H_2/Ar at 600°C for 4 h, C) after exposure to air for 1 min and D) after reoxidation in 10% O_2/Ar at 600°C for 4 h overlaid with S) reference spectrum of a NiO foil.

BE of 852.2 eV, spectrum not shown), except the core line was shifted 1 eV lower, to 851.2 eV. The negative BE shift, noted for Ni metal, may be a consequence of a strong interaction between metal clusters and the insulating support (*26*). Upon exposure to air (Figure 3C), peaks appeared at 853.8, 855.3, and 860.9 eV. These three lines match that of the NiO standard. After reoxidation at 600°C, the line for Ni^0 disappeared, and the remaining core envelope resembled a combination of Ni^{2+} and "Ni^{3+}" states.

In a previous XPS study of NiHZSM-5 catalysts, Badrinarayanan *et al.* (*3*) concluded that nickel was present always as Ni^{2+}, even after exposure to harsh reducing conditions. As previously revealed, cobalt exchanged within the zeolite pores was in the +2 oxidation state. The stable oxidation state of nickel is +2. Thus, it seems logical that if nickel occupied zeolite framework sites, it would be present predominantly in the +2 oxidation state. Indeed, the Ni/(Al-Na) atomic ratio measured by ICP-AES after the initial calcination (500°C in air) was 1.1 (Table I), suggesting that nickel probably occupied interstitial sites. However, nickel was reduced completely in hydrogen, implying that the nickel probed by XPS was a surface species. This is substantiated by data in Table II; after *in situ* oxidation, the surface Ni/Si ratio was 6 times higher than that measured for the bulk, i.e. 13×10^{-2} vs. 2.2×10^{-2}.

Collectively, these results for the Ni-containing sample indicate that while initially Ni^{2+}-ions may have exchanged into the ZSM-5 they are not very stable in the structure and, contrary to Co^{2+}-ions, migrate out to the surface during high temperature (600°C) *in situ* oxidative treatment and become susceptible to reduction by hydrogen. The XPS Ni/Si ratio after reduction was lowered to 3.2×10^{-2}, implying that the surface nickel agglomerated to form larger crystallites, only the top of which are probed by XPS.

The higher BE species initially noticed for nickel after oxidation (Figure 3A) may result from an interaction between highly dispersed nickel and the zeolite surface. Wu and Hercules (*27*) studied nickel supported catalysts on γ-Al_2O_3 and noticed that NiO was not present at low loadings, where nickel was dispersed and highly interacting with the support. Instead Wu and Hercules observed a Ni 2p spectrum essentially identical to that reported by Kim and Davis (*25*) for Ni_2O_3. However, Wu and Hercules were careful not to identify the spectrum as owing to Ni^{3+}, and instead qualified their results as due to an interaction between nickel and the γ-Al_2O_3 substrate. It is interesting to note though, that while higher BEs for dispersed metal oxides on γ-Al_2O_3 not uncommon, in the case of cobalt the spectral envelope changed also. Thus, while the Ni 2p spectrum for NiO shows one main line and two satellites, the "highly interacting" nickel species reveals a spectrum with only one satellite line, and looks just like Ni^{3+}, which strongly suggests that the a Ni^{3+} species may indeed be present.

Note that after reoxidation, some of the nickel appears to have remained as Ni^{2+}. Mosty likely, surface crystals of NiO were formed from nickel metal clusters present at the external surface (formed during reduction).

Cerium-exchanged NaZSM-5. XPS spectra acquired following oxidation, reduction, exposure to air, and then reoxidation of the cerium-exchanged zeolite are given

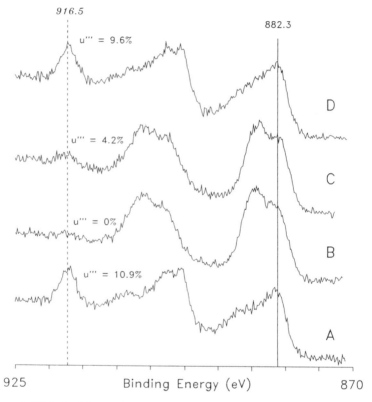

Figure 4. XPS Ce 3d core level spectra of CeNaZSM-5 A) after oxidation in 10% O$_2$/Ar at 600°C for 4 h, B) after reduction in 10% H$_2$/Ar at 600°C for 4 h, C) after exposure to air for 1 min and D) after reoxidation in 10% O$_2$/Ar at 600°C for 4 h.

in Figure 4. Ceria contains Ce-ions either in the +3 or +4 oxidation state. The main 3d core line BE shift between the +3 and +4 states is minimal, and cannot be used to differentiate the two. However, the XPS 3d core level spectrum of Ce^{4+} includes a unique and unobscured peak, designated as u''', which can be used to discriminate Ce^{4+} from Ce^{3+} (*14*). The contribution of the u''' peak to the total Ce 3d signal area has been shown to vary linearly with Ce^{4+} content (*28*). Total oxidation to Ce^{4+} gives a u''' peak of 13.4% of the total Ce 3d envelope.

Figure 4A demonstrates that after oxidation at 600°C, 81% (10.9/13.4 × 100) of the Ce was in the +4 state. The remainder of the ceria (Ce^{3+}) which was nonsusceptible to oxidation must have been stabilized through an interaction with either the surface or the inner framework of the zeolite. Figure 4B shows that all ceria was present as +3 after reduction. Upon exposure to air (Figure 4C), about 31% of the ceria was reconverted to Ce^{4+}. The high temperature reoxidation (Figure 4D) converted slightly less ceria to +4 (72%) than was noticed in the original oxidation. Some surface enrichment is observed for the Ce-ions exchanged into the NaZSM-5 (Table II).

Lanthanum-exchanged NaZSM-5. The XPS 3d core level spectrum of the lanthanum-exchanged zeolite after oxidative treatment overlaid with a spectrum of dispersed lanthanum on γ-Al_2O_3 (*29*), also oxidized at 600°C, is shown in Figure 5. The spectra are essentially identical, exhibiting a La $3d_{5/2}$ BE of 835.2 eV. The La $3d_{5/2}$ BE reported for particulate phase La_2O_3 is 833.2 eV (*29*). The shift to higher BE noticed for dispersed phase lanthana on γ-Al_2O_3 was attributed to a low coordination of La^{3+} ions in intimate contact with the alumina support. No change in the La 3d core level occurred after reduction at 600°C in H_2, as was expected for valence invariant La^{3+} (*14*). The surface enrichment of the La-ions is somewhat lower than that of Ce- or Fe-ions (Table II).

Palladium-exchanged NaZSM-5. XPS 3d spectra acquired after oxidation and reduction of the palladium-exchanged zeolite are given in Figure 6. The Pd $3d_{5/2}$ BE after oxidation was 336.6 eV, essentially identical to that observed for a reference PdO foil (*30*). The Pd/Si atomic ratio measured by XPS was 9.1×10^{-2} (Table II), considerably higher than the bulk ratio of 2.5×10^{-2}. Clearly, the palladium observed by XPS was at the external surface and migrated there at high temperature even under oxidizing conditions. This palladium was completely reduced to the metallic state in hydrogen (Figure 6B). The BE measured after reduction was 334.5 eV, which is about 0.5 eV lower than expected for bulk Pd^0 (*30*). A negative BE shift was also noted for nickel in NiNaZSM-5 after reduction. As was surmised for the nickel zeolite, the lower BE shift of Pd metal may result from an interaction between the metal and insulating support.

Copper-exchanged ZSM-5 zeolites. Copper was exchanged into HZSM-5, NaZSM-5, and LaNaZSM-5 zeolites. The XPS Cu $2p_{3/2}$ core level spectra acquired after oxidative and reductive treatments are shown in Figure 7. Similar spectra were obtained for each zeolite after oxidation. The Cu $2p_{3/2}$ envelope featured three prevalent peaks, two main lines at 933.5 and 936.1 eV, and a satellite peak (which

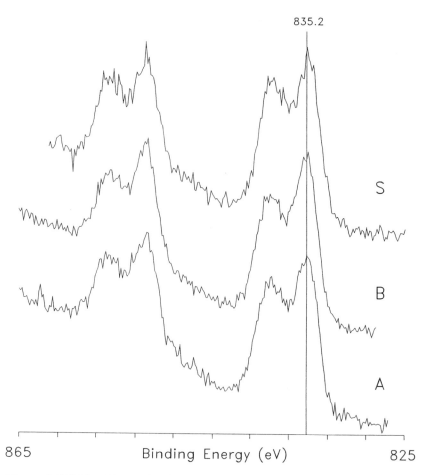

Figure 5. XPS La 3d core level spectra of LaNaZSM-5 A) after oxidation in 10% O$_2$/Ar at 600°C for 4 h and B) after reduction in 10% H$_2$/Ar at 600°C for 4 h overlaid with S) reference spectrum of dispersed La/γ-Al$_2$O$_3$ (La/Al atomic ratio = 0.05).

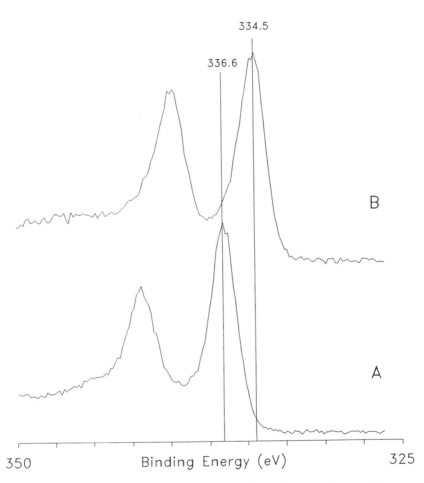

Figure 6. XPS Pd 3d core level spectra of PdNaZSM-5 A) after oxidation in 10% O_2/Ar at 600°C for 4 h and B) after reduction in 10% H_2/Ar at 600°C for 4 h.

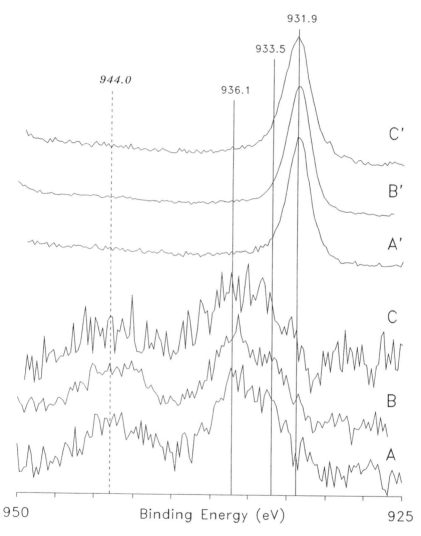

Figure 7. XPS Cu 2p core level spectra of A) CuHZSM-5, B) CuNaZSM-5 and C) CuLaNaZSM-5 after oxidation in 10% O$_2$/Ar at 600°C for 4 h and A') CuHZSM-5, B') CuNaZSM-5 and C') CuLaNaZSM-5 after reduction in 10% H$_2$/Ar at 600°C for 4 h.

was a combination of unresolvable peaks) centered at 944.0 eV. The higher BE core line at 936.1 eV was the dominant peak in all three spectra. The Cu spectra are the same as previously recorded for copper exchanged HZSM-5 (*31*). Both core level peaks are assigned to the Cu^{2+} state. The lower core line is similar in BE to CuO, while the higher BE line is associated with a chemical state highly ionic in nature. It is noteworthy that the other ions present in the ZSM-5 (H^+, Na^+, or La^{3+}-Na^+) did not affect the shape of the copper spectra.

The XPS Cu/Si ratio observed for CuLaNaZSM-5 was only a quarter of that measured for CuNaZSM-5 while the difference in the bulk content was only half as large (Table II). Putting this in different terms one notes that, after high-temperature calcination, similarly to cobalt the copper ions are uniformly distributed in the zeolite. In the presence of lanthanum the surface copper is masked by the Co-ion. This is apparent in the poor signal intensity obtained during acquisition of the Cu $2p_{3/2}$ spectrum (Figure 7C). Unfortunately, acquisition times for Cu spectra are time-limited, since reduction of Cu^{2+} takes place during prolonged exposure to the X-ray beam (*31*).

The corresponding Cu $2p_{3/2}$ spectra for the copper exchanged zeolites obtained after hydrogen reduction were all identical (Figures 7A', 7B' and 7C') . The satellite peak observed for the oxidized samples had disappeared, and one core line was observed at a BE of 931.9 eV. The spectra match that of either Cu^0 or Cu^{1+} (*31*). However, reduction with H_2 at 600°C is expected to reduce copper completely to the metallic state (*31*). The presence of lanthanum does not protect the copper ions from reduction by hydrogen which is also corroborated by the surface enrichment of Cu during reduction (Table II).

Conclusions

The examination of a series of NaZSM-5 zeolites exchanged by a series of divalent and trivalent metal ions after high-temperature treatments in O_2 and H_2 revealed important differences in the distribution ot these ions within the zeolites, their valence states and resistance to migration out of the zeolitic framework. In general the behavior parallels that noted by Minachev *et al.* (*32*) who have reported that the mobility of divalent metals in zeolites decreases in the order: Pd > Cu > Ni >> Co. In our case the relative order of Cu and Ni is reversed. Trivalent ions with higher oxidation potentials are not distributed uniformly in the lattice but are less susceptible to outmigration in strongly reducing conditions.

Literature Cited

1. Ono, Y. *Catal. Rev. Sci. Eng.* **1992**, *34*, 179.
2. Hamid, S. B. A.; Derouane, E. G.; Demortier, G.; Riga, J.; Yarmo, M. A. *Appl. Catal.* **1994**, *108*, 85.
3. Badrinarayanan, S.; Hedge, R. I.; Balakrishnan, I.; Kulkarni, S. B.; Ratnasamy, P. *J. Catal.* **1981**, *71*, 439.
4. Rao, V. U. S. *Physica Scripta.* **1983**, *T4*, 71-8.
5. Iwamoto, M.; Hamada, H. *Catal. Today* **1991**, *10*, 57.

6. Truex, T.J.; Searles, R. A.; Sun, D. C. *Plat. Met. Rev.* **1992**, *36*, 2.
7. Yokoyama, C.; Misono, M. *Chem. Lett.* **1992**, 1669.
8. Kikuchi, E.; Terasaki, I.; Ihara, M.; Yogo, K. *Preprints Div. Petr. Chem.* 207th National Meeting ACS, San Diego, CA, March 13-18, **1994**, 160.
9. Sano, T.; Suzuki, K.; Shoji, H.; Ikai, S.; Okabe, K.; Murakami, T.; Shin, S.; Hagiwara, H.; Takaya, H. *Chem. Lett.* **1987**, 1421.
10. Suzuki, K.; Sano, T.; Shoji, H.; Murakami, T.; Ikai, S.; Shin, S.; Hagiwara, H.; Takaya, H. *Chem. Lett.* **1987**, 1507.
11. Grinsted, R. A.; Jen, H.-W.; Montreuil, C. N.; Rokosz, M. J.; Shelef, M. *Zeolites* **1993**, *13*, 602.
12. Bryson III, C. E. *Surf. Sci.* **1987**, *189-190*, 50.
13. Scofield, J. H. *J. Electron Spectrosc. Relat. Phenom.* **1976**, *8*, 129.
14. Shelef, M.; Haack, L. P.; Soltis, R. E.; deVries, J. E.; Logothetis, E. M. *J. Catal.* **1992**, *137*, 114.
15. McInytre, N. S.; Zetaruk, D. G. *Anal. Chem.* **1977**, *49*, 1521.
16. Yue, Y,; Shen, W.; Yen, Y.; Ding, Y. *Gaodeng Xuexiao Huaxue Xuebao.* **1992**, *13*, 1503.
17. Wagner, C. D.; Riggs, W. M.; Davis, L. E.; Moulder, J. F. In *Handbook of X-ray Photoelectron Spectroscopy*; Muilenbery, G. E., Ed.; 1st Edition; Perkin-Elmer Corporation, Physical Electronics Division: Eden Prairie, MN, 1979, p 76.
18. Chuang, T. J.; Brundle, C. R.; Rice, D. W. *Surf. Sci.* **1976**, *59*, 413.
19. McIntyre, N. S.; Cook, M. G. *Anal. Chem.* **1975**, *47*, 2208.
20. Chin, R. L.; Hercules, D. M. *J. Phys. Chem.* **1982**, *86*, 360.
21. Wei, Q.; Ying, C.; *Gaodeng Xuexiao Huaxue Xuebao.* **1991**, *12*, 80.
22. Stencel, J. M.; Rao, V. U. S.; Diehl, J. R.; Rhee, K. H.; Dhere, A. G.; DeAngelis, R. J. *J. Catal.* **1983**, *84*, 109.
23. Rossin, J. A.; Saldarriaga, C.; Davis, M. E. *Zeolites* **1987**, *7*, 295.
24. Wagner, C. D.; Riggs, W. M.; Davis, L. E.; Moulder, J. F. In *Handbook of X-ray Photoelectron Spectroscopy*; Muilenbery, G. E., Ed.; 1st Edition; Perkin-Elmer Corporation, Physical Electronics Division: Eden Prairie, MN, 1979, p 78.
25. Kim, K. S.; Davis, R. E. *J. Electron Spectrosc.* **1972**, *1*, 251.
26. Mason, M. G. *Phys. Rev. B* **1983**, *27*, 748.
27. Wu, M.; Hercules, D. M. *J. Phys. Chem.* **1979**, *83*, 2003.
28. Shyu, J. Z.; Otto, K.; Watkins, W. L. H.; Graham, G. W.; Belitz, R. K.; Gandhi, H. S. *J. Catal.* **1988**, *114*, 23.
29. Haack, L. P.; deVries, J. E.; Otto, K.; Chatta, M. S. *Appl. Catal. A* **1992**, *82*, 199.
30. Otto, K.; Haack, L. P.; deVries, J. E. *Appl. Catal. B* **1992**, *1*, 1.
31. Haack, L. P.; Shelef, M. in *Chap. 6, ACS Sympos. Ser., Environ. Catal.* Ed: Armor, J. N., Amer. Chem. Societ., Wash. D. C., **1993**, pp. 66-73.
32. Minachev, Kh. M.; Antoshin, G. V.; Shpiro, E. S.; Yusifov, Yu. A. in *Proc. Int. Congr. Catal.*; Bond, G. C.; Wells, P. B.; Tompkins, F. C., Eds.; 6th; Chem. Soc.: Letchworth, Engl., 1976, Vol 2, pp 621-32.

RECEIVED October 31, 1994

Chapter 14

NO+CO Reaction on Rh and CeRh–SiO$_2$ Catalysts

In Situ IR and Temperature-Programmed Reaction Study

Steven S. C. Chuang, Raja Krishnamurthy, and Girish Srinivas

Department of Chemical Engineering, University of Akron,
Akron, OH 44325-3906

Interaction and reaction of adsorbed NO and CO on Rh and Ce-Rh catalysts have been studied by combined *in situ* infrared spectroscopy and temperature-programmed reaction at 298-673 K. At 298 K, adsorption of NO as NO$^-$ causes the desorption of preadsorbed linear and bridged CO from reduced Rh sites. NO adsorption competes over CO adsorption on the reduced Rh while NO does not adsorb on oxidized Rh catalyst which chemisorbs CO as gem-dicarbonyl. Temperature-programmed reaction study reveals that the type of adsorbate and the surface state of the catalyst change with temperature. At 453 - 543 K, NO adsorbs as low wavenumber NO$^-$ at 1689-1696 cm^{-1} which may be involved in NO dissociation and CO adsorbs as gem-dicarbonyl which may be a spectator species for this reaction.

The largest application of Rh as a catalyst is in the automobile catalytic converter because of its unique activity for reduction of NO$_x$ and the oxidation of CO and hydrocarbons (*1*). The scarcity and high price of Rh and increasingly stringent standards for NO$_x$ emissions have prompted extensive studies to further improve the performance and to develop substitutes for Rh-based catalysts. Improvement of Rh performance for NO and CO reaction lies in our understanding of the reactivity of adsorbate, the nature of active sites, and the reaction pathway. Several previous studies have suggested that the reaction pathway for CO$_2$ formation involves the reaction of adsorbed CO with adsorbed O produced from the dissociation of NO; adsorbed N atoms from NO dissociation combine to form N$_2$ (*1-4*).

Although the reaction pathway for the formation of N$_2$ and CO$_2$ has been elucidated, the type of adsorbed NO and CO involved in the reaction remains unknown. The types of CO and NO adsorbed on Rh are closely related to the surface state of Rh and the partial pressure of NO and CO (*5-12*). An investigation of how the various types of adsorbed CO and NO interact and react may provide some insight into the nature of Rh sites that are active for the NO and CO reaction.

The objectives of this study are (i) to investigate the interaction between adsorbed CO and NO by exposing adsorbed CO to gaseous NO and adsorbed NO to gaseous CO at 298 K and (ii) to study the nature of adsorbed CO and NO over Rh/SiO$_2$ and

0097–6156/95/0587–0183$12.00/0

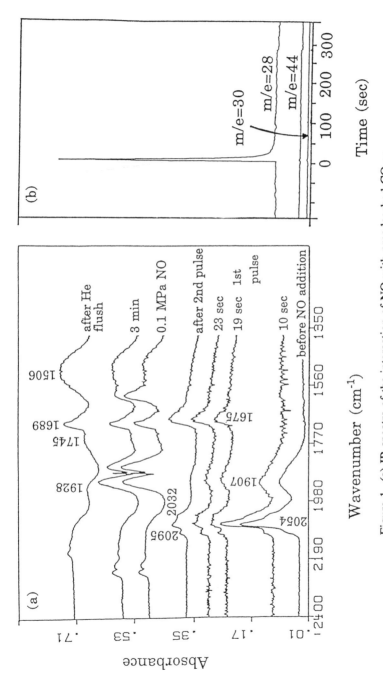

Figure 1. (a) IR spectra of the interaction of NO with preadsorbed CO on reduced Rh/SiO$_2$ at 298 K.
(b) MS analysis of the effluent from the IR cell for Figure 1.

Ce-Rh/SiO$_2$ catalysts during temperature-programmed reaction of NO and CO from 298 to 673 K.

Experimental

A 4 wt% Rh/SiO$_2$ catalyst was prepared by incipient wetness impregnation of large pore silica support (Strem, 350 m^2/g) using RhCl$_3$•3H$_2$O (Alfa Chemicals) solution. The Ce-Rh/SiO$_2$ catalyst with 5 wt% Rh and 5 wt% Ce was prepared by co-impregnating a solution of RhCl$_3$•3H$_2$O and Ce(NO$_3$)$_3$•6H$_2$O (Alfa) onto the silica support. The ratio of the solution to the weight of support material used is 1 cm^3 to 1 gm. The catalysts were dried in air overnight at 303 K followed by reduction in flowing hydrogen at 673 K for 8 hrs.

An infrared (IR) reactor cell capable of operation upto 873 K and 6 MPa was used for this study (*13-15*). The catalysts were pressed into self-supporting disks and loaded into the IR reactor cell. Prior to the experiments, the catalysts were further reduced in H$_2$ flow or oxidized in flowing air in the IR reactor cell at 573 K for 2 hrs. Adsorption of NO and CO on the catalyst surface was carried out by flowing 0.1 MPa of CO or NO into the reactor (*16*). 15 μl of NO was pulsed over preadsorbed CO to study the interaction of adsorbed CO with gaseous NO. Pulse CO chemisorption study at 303 K showed that the Rh/SiO$_2$ catalyst chemisorbed 55.3 μmol CO/g catalyst and the Ce-Rh/SiO$_2$ catalyst chemisorbed 28.5 μmol CO/g catalyst (*13*). Apparent Rh crystallite sizes were estimated to be 63^{0}A for Rh/SiO$_2$ and 154^{0}A for Ce-Rh/SiO$_2$ assuming CO$_{ad}$/Rh=1. It should be noted that Ce may cover part of the Rh surface affecting CO chemisorption. The large Rh crystallite size is the result of direct reduction of RhCl$_3$ without prior calcination and the use of high loading of Rh. Infrared spectra of adsorbates were obtained by a Fourier transform infrared (FTIR) spectrometer at a resolution of 4 cm^{-1}. The effluent from the IR cell was monitored by a Balzers QMG 112 quadrupole mass spectrometer.

Results and Discussion

Interaction of NO with Preadsorbed CO on Rh/SiO$_2$, Oxidized Rh/SiO$_2$, and Ce-Rh/SiO$_2$. Figure 1(a) shows the infrared spectra of the interaction of NO with preadsorbed CO on the reduced Rh/SiO$_2$ catalyst at 298 K. CO chemisorption on the reduced Rh/SiO$_2$ at 0.1 MPa of CO and 298 K followed by removal of gaseous CO by flushing with He produced a linear CO band at 2054 cm^{-1} and a bridged CO band at 1907 cm^{-1}. Transient infrared spectra during the first NO pulse showed a higher rate of decrease in the intensity of the linear CO band than that of the bridged CO band indicating that the linear CO was displaced more rapidly than the bridged CO by gaseous NO. The NO pulse also led to the development of an NO$^-$ band at 1675 cm^{-1}. It should be noted that transient IR spectra were taken with 4 coadded scans which gave a low signal to noise ratio. The spectra obtained after the second NO pulse showed little change. Figure 1(b) shows the response of the effluent from the IR reactor during the first NO pulse. CO, NO, and CO$_2$ correspond to the m/e ratios 28, 30, and 44, respectively; however, NO-CO reaction products such as N$_2$ and N$_2$O could also contribute to the m/e ratios of 28 and 44, respectively. The intensity of a minute m/e=44 peak during the NO pulse over the catalyst was the same as that of the m/e=44 peak during the blank run suggesting that both CO$_2$ and N$_2$O were not formed on the Rh/SiO$_2$ catalyst at 298 K. The blank run is performed by direct injection of the inlet reactants into the mass spectrometer. The minute m/e=44 peak obtained from the blank run is due to contamination in the reactant gas stream. The reaction

$$NO(g) + CO_{ad} \rightarrow CO_2 + \frac{1}{2}N_2$$

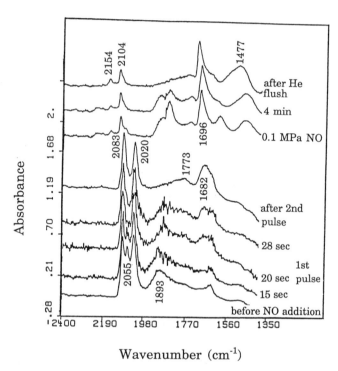

Figure 2. IR spectra of the interaction of NO with preadsorbed CO on oxidized Rh/SiO$_2$ at 298 K.

did not occur. All the 15 μl of NO in the first pulse was adsorbed as NO⁻ as indicated by the absence of NO breakthrough in Figure 1(b). Flowing 0.1 MPa of NO over the catalyst resulted in further depletion of linear and bridged CO and the development of additional bands at 1928 cm⁻¹ due to Rh-NO⁺, 1745 cm⁻¹ due to Rh-NO⁻, and 1506 cm⁻¹ due to bidentate nitrato species. Gas phase N_2O bands at 2230 and 2204 cm⁻¹ were also observed. The assignments of these bands agree well with those reported in literature (*2,9,10,13,17*). The gas phase NO and N_2O bands were removed by flushing with He and little change in the adsorbed NO bands was observed. Interaction of adsorbed CO with gaseous NO did not result in the formation of CO_2 at 373 and 423 K (*13*). The lack of CO_2 formation could be due to the desorption of adsorbed CO before dissociation of NO.

Figure 2 shows the IR spectra of the interaction of NO with preadsorbed CO on the oxidized Rh/SiO$_2$ at 298 K. CO chemisorption on the oxidized Rh/SiO$_2$ at 0.1 MPa of CO and 298 K followed by removal of gaseous CO by flushing with He produced a linear CO band at 2055 cm⁻¹, a bridged CO band at 1893 cm⁻¹, and gem-dicarbonyl bands at 2083 and 2020 cm⁻¹. The presence of linear and bridged CO could be either due to the incomplete oxidation of Rh surface in 2 h at 573 K or due to the partial reduction of Rh sites by CO. Transient infrared spectra during the first NO pulse showed a gradual decrease in the intensity of the linear CO and bridged CO bands with the development a NO⁻ band at 1682 cm⁻¹. The rate of decrease of linear CO appears to be higher than that of bridged CO. The spectra obtained after the second NO pulse showed little change except for the formation of a weak adsorption band at 1773 cm⁻¹ due to Rh-NO⁻ species. Introduction of 0.1 MPa of NO removed gem-dicarbonyl bands and produced bidentate nitrato species at 1477 cm⁻¹, adsorbed NO⁻ at 1696 cm⁻¹, and a cyanide (CN) species at 2154 cm⁻¹. The introduction of NO caused the increase in intensity and wavenumber of adsorbed NO⁻. Removal of gas phase NO caused a further upward shift in wavenumber of adsorbed NO⁻. The amount of NO adsorbed was found to be less on the oxidized catalyst than on the reduced Rh/SiO$_2$.

Figure 3 shows the IR spectra of the interaction of NO with preadsorbed CO on Ce-Rh/SiO$_2$ catalyst at 298 K. CO adsorption produced linear CO at 2055 cm⁻¹, bridged CO at 1865 cm⁻¹, and a band at 1738 cm⁻¹ assigned to tilted CO (*13*). Transient infrared spectra during the first NO pulse caused the decrease in the intensity of linear CO band and the development of a band due to NO⁻ at 1682 cm⁻¹. The second NO pulse resulted in the further decrease in intensity of linear and bridged CO, increase in the NO⁻ intensity, and the emergence of a weak band centered around 2020 cm⁻¹. Introduction of 0.1 MPa of NO produced adsorbed cyanide species at 2154 cm⁻¹, bidentate nitrato species at 1513 cm⁻¹, and a slight increase in the intensity of NO⁻ band at 1682 cm⁻¹. The desorption of linear and bridged CO resulting from NO adsorption as NO⁻ species suggests that NO⁻ may be adsorbed on the reduced Rh crystallite surface which chemisorbs linear and bridged CO. The results also indicate that adsorbed NO⁻ is more strongly bonded than adsorbed CO on these reduced Rh sites.

Interaction of CO with Preadsorbed NO on Rh/SiO$_2$, Oxidized Rh/SiO$_2$, and Ce-Rh/SiO$_2$. The infrared spectra of the interaction of CO with preadsorbed NO on the reduced Rh/SiO$_2$ at 298 K is shown in Figure 4. Exposure of the catalyst to NO produced adsorbed NO⁻ bands at 1689 and 1738 cm⁻¹, adsorbed NO⁺ at 1921 cm⁻¹, and a bidentate nitrato band at 1527 cm⁻¹. The IR bands remained unchanged upon pulsing 15 μl of CO over the catalyst surface. Introduction of 0.1 MPa of CO into the reactor did not result in any changes in the adsorbed bands. Removal of gaseous CO by He flush revealed the presence of weak gem-dicarbonyl bands and a cyanide band at 2160 cm⁻¹. Significantly more gem-dicarbonyl and cyanide species were formed at 373 K than at 298 K (*13*). The formation of gem-dicarbonyl and cyanide species suggests that preadsorbed NO oxidizes the Rh surface leaving adsorbed N to

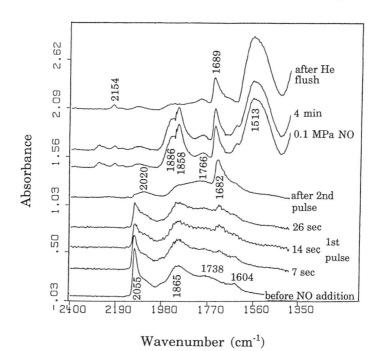

Figure 3. IR spectra of the interaction of NO with preadsorbed CO on Ce-Rh/SiO$_2$ at 298 K.

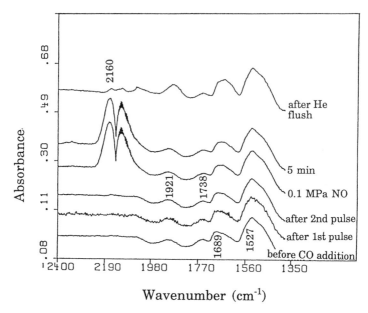

Figure 4. IR spectra of the interaction of CO with preadsorbed NO on reduced Rh/SiO$_2$ at 298 K.

further react with dissociated CO to form cyanide species. Analysis of the reactor effluent did not reveal the formation of CO$_2$ product. Similar results were obtained for Ce-Rh/SiO$_2$ catalyst. NO did not adsorb on the oxidized Rh catalyst indicating that dissociative adsorption of NO may occur only on the reduced Rh sites.

Temperature-Programmed Reaction (TPR). TPR under an NO-CO-He (NO:CO:He=1:1:3) environment was undertaken at a heating rate of 10 K/min from 298 K to 673 K in the reactor cell. Helium is an inert gas in the reaction and was used to keep the concentration of ionized species in the MS vacuum chamber low for obtaining high sensitivity to measurement of reaction products. The catalyst was exposed to NO-CO-He flow at 573 K for 6 hr followed by reduction at 573 K for 2 hr and cooled under H$_2$, prior to the TPR. The NO-CO reaction at 573 K resulted in the development of the isocyanate species on SiO$_2$ which cannot be removed from the surface by hydrogen reduction at 573 K. Figure 5 shows the product response during the temperature-programmed NO-CO reaction on Rh/SiO$_2$. Since both CO and N$_2$ give m/e at 28 and both CO$_2$ and N$_2$O yield m/e=44, the m/e=28 and 44 profiles provide only qualitative information for CO conversion and N$_2$ formation. The profiles of CO and N$_2$ (m/e=28) and NO (m/e=30) showed a gradual decrease with a corresponding gradual increase in the CO$_2$ and N$_2$O (m/e=44) profiles. Sudden changes in the profiles of the species was observed at 530 K from the reaction which showed a dramatic increase in conversion of NO and formation of CO$_2$ and NO$_2$.

Figure 6 shows the IR spectra as a function of temperature during the TPR on Rh/SiO$_2$ catalyst. The spectrum of the catalyst surface obtained in the absence of NO and CO was used as the background spectrum which is dependent only on the temperature. The spectrum obtained at 303 K shows the following: (i) a Si-NCO band at 2309 cm^{-1}, (ii) an N$_2$O band at 2241 cm^{-1}, (iii) gaseous CO bands at 2166 and 2116 cm^{-1}, (iv) gaseous NO band at 1916 cm^{-1}, (v) an adsorbed NO band at 1863 cm^{-1}, (vi) adsorbed NO$^-$ bands at 1759 and 1695 cm^{-1}, and (vii) a bidentate nitrato species at 1501 cm^{-1}. The assignment of these bands are consistent with literature results (*1,5-17*). This spectrum under a NO-CO environment was significantly different from that observed under either pulsing NO over preadsorbed CO or pulsing CO over preadsorbed NO environments at 298 K. The partial pressure of NO and CO affects the adsorbed species and the surface state of the catalyst. Increasing the temperature to 453 K resulted in the disappearance of the shoulder at 2343 cm^{-1}, the Rh-NO band at 1863 cm^{-1}, the band at 1621 cm^{-1}, and the bidentate nitrato species at 1501 cm^{-1} and emergence of gem-dicarbonyl bands at 2098 and 2032 cm^{-1} and Rh$^+$(CO) band at 2105 cm^{-1}.

As temperature is increased from 453 to 543 K, the following spectral changes were observed: (i) appearance of a weak CO$_2$ band at 2358 cm^{-1}, (ii) appearance of Rh-NCO at 2189 cm^{-1}, (iii) disappearance of the bidentate nitrato species, (iv) an increase in gem-dicarbonyl bands followed by a gradual decrease, (v) gradual formation of Rh-NO$^+$ band at 1913 cm^{-1}, and (vi) sudden increase in the intensity of NO$^-$ band at 1695 cm^{-1} and disappearance of NO$^-$ band at 1759 cm^{-1}.

The high wavenumber (1740-1770 cm^{-1}) NO$^-$ at 1759 cm^{-1} was observed from 303 to 453 K while the low wavenumber (1650-1700 cm^{-1}) NO$^-$ at 1695 cm^{-1} became prominent at temperatures between 473 and 543 K. The low NO stretching frequency for NO$^-$ has been suggested to be an indication of the weakening of the N-O bond and the strengthening of the of the Rh-N bond compared with those of neutral NO at 1830 cm^{-1} and NO$^+$ at 1910 cm^{-1} (*2,8*). Accordingly, the low wavenumber NO$^-$ at 1695 cm^{-1} should be more strongly bonded to the Rh than the high wavenumber NO$^-$ at 1759 cm^{-1}. The low wavenumber NO$^-$ on the reduced Rh crystallite may be the precursor for NO dissociation. Large Rh crystallites have also been reported to favor NO dissociation (*18*).

The formation of the gem-dicarbonyl and NO$^+$ in the temperature range between 453 and 543 K indicates that part of Rh surface has been oxidized by NO resulting in

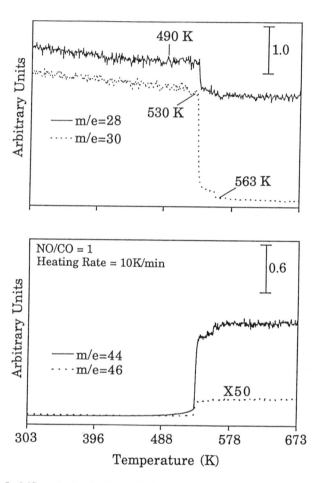

Figure 5. MS analysis of effluent during temperature-programmed NO-CO reaction on Rh/SiO$_2$. Heating rate=10K/min (NO:CO:He=1:1:3).

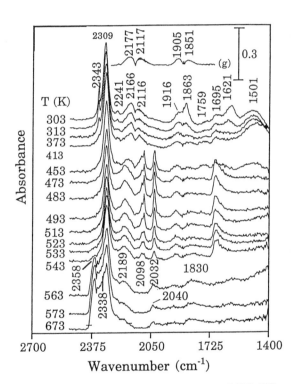

Figure 6. IR spectra during temperature-programmed NO-CO reaction on Rh/SiO₂.

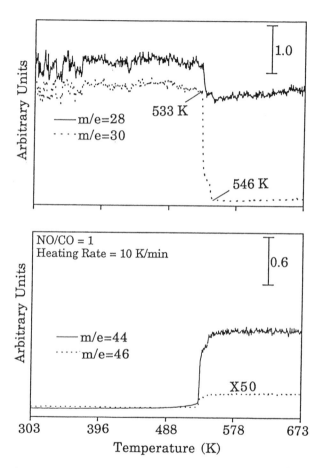

Figure 7. MS analysis of effluents during temperature-programmed NO-CO reaction on Ce-Rh/SiO$_2$. Heating rate=10K/min (NO:CO:He=1:1:3).

the formation of Rh$^+$ sites that chemisorb gem-dicarbonyl and NO$^+$. The role of gem-dicarbonyl in the formation of CO_2 during NO+CO reaction remains unclear. Gem-dicarbonyl on Rh$^+$ sites has never been observed on the Rh single crystal (*19*) which has been demonstrated to be very active for NO and CO reaction (*4*). The gem-dicarbonyl showed little reactivity towards 0.1 MPa of O_2 on Rh/SiO$_2$ at 358 K (*20*). Temperature above 448 K is required for the removal of the adsorbed oxygen by reductive agglomeration to form CO_2 and linear CO from the gem-dicarbonyl in the presence of 50 torr CO (*21*). The reductive agglomeration appears to depend upon the partial pressure of CO and NO. Gem-dicarbonyl band was completely removed and a weak linear CO band emerged as temperature increased above 543 K. The weak bands for adsorbed NO and linear CO observed at temperatures above 543 K could be due to mass transfer limitations occurring under these reaction conditions where the diffusion of the reactants may not be fast enough to provide reactants for the reaction. Activation energy between 473 and 573 K was determined to be 9 Kcal/mole which is significantly lower than the value (33 Kcal/mole) reported in the literature which suggests that the reaction is internal and external diffusion limited. No changes were observed in the IR spectra upon increasing the temperature to 673 K.

The profiles of CO and N$_2$ (m/e=28), NO (m/e=30), N$_2$O and CO_2 (m/e=44), and NO$_2$ (m/e=46) during NO-CO TPR on Ce-Rh/SiO$_2$ catalyst are shown in Figure 7. The profiles are similar to those observed for Rh/SiO$_2$. The temperature for 99% conversion on Ce-Rh/SiO$_2$ is lowered by 17 K to 546 K than that on Rh/SiO$_2$. Figure 8 shows the IR spectra during the TPR of NO-CO on Ce-Rh/SiO$_2$ catalyst. The spectrum at 303 K showed the Si-NCO band at 2309 cm^{-1}, a weak N$_2$O band at 2241 cm^{-1}, gaseous CO bands at 2166 and 2116 cm^{-1}, gaseous NO band at 1916 cm^{-1}, adsorbed NO band at 1863 cm^{-1}, NO$^-$ bands at 1763 and 1691 cm^{-1}, and a bidentate nitrato band at 1502 cm^{-1}.

An increase in temperature to 413 K resulted in decrease in intensity of NO band at 1863 cm^{-1} and NO$^-$ band at 1621 cm^{-1} and a slight increase in the 1763 cm^{-1} band. A distinct band at 2157 cm^{-1} assigned to a cyanide species and a shoulder band at 2189 cm^{-1} due to Rh-NCO developed. Further increase in temperature to 543 K resulted in (i) a gradual increase in the Rh-NCO band followed by a slight decrease, (ii) disappearance of cyanide band, (iii) formation of gem-dicarbonyl bands at 2098 and 2032 cm^{-1} followed by a decrease in their intensity, (iv) decrease in the intensity of NO$^-$ bands at 1763 and 1691 cm^{-1} and bidentate nitrato species at 1502 cm^{-1}, and (v) the formation of NO$^+$ band at 1916 cm^{-1}. A weak gas phase CO_2 also developed at 2358 cm^{-1}. Increase in temperature above 543 K resulted in significant decrease in NO$^-$ band, the disappearance of gem-dicarbonyl bands, the depletion of Rh-NCO and bidentate nitrato bands, and the formation of strong CO_2 bands at 2358 and 2338 cm^{-1}.

A comparison of TPR results on Rh/SiO$_2$ and Ce-Rh/SiO$_2$ shows (i) Ce decreases the temperature for 99% conversion, (ii) Ce decreases the spillover of isocyanate from Rh to SiO$_2$, (iii) Ce stabilizes the formation of bidentate nitrato species at temperatures above 473 K, (iv) Ce decreases the formation of N$_2$O as indicated by a decrease in intensity of 2240 cm^{-1} band and decreases the formation of NO$_2$ as indicated by the MS response over Ce-Rh/SiO$_2$ catalyst, and (v) Ce decreases the formation of Rh$^+$(CO) which exhibits a shoulder band at 2105 cm^{-1}.

Conclusions

The key results of interaction and reaction of adsorbed CO and NO are summarized below

(i) Adsorption of NO as NO$^-$ causes the desorption of linear and bridged CO from reduced Rh sites at 298 K.

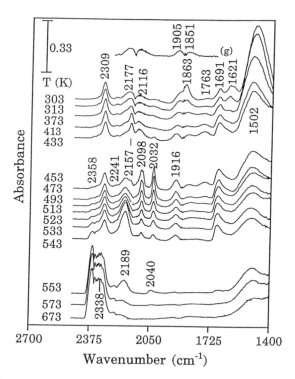

Figure 8. IR spectra during temperature-programmed NO-CO reaction on Ce-Rh/SiO$_2$.

(ii) Adsorption of NO on the reduced Rh causes the oxidation of Rh to Rh$^+$ which can chemisorb CO as gem-dicarbonyl and NO as NO$^+$. NO does not adsorb on the oxidized Rh surface at 298 K.

(iii) The type of adsorbate and the surface state of catalyst change with temperature during NO-CO reaction.

(iv) Between 453 and 543 K, gem-dicarbonyl is the major adsorbed species which may be a spectator. Its role in the reaction will be further investigated.

(v) At temperatures higher than 543 K, where NO conversion is essentially 99%, reductive agglomeration of the Rh surface takes place indicated by the disappearance of gem-dicarbonyl CO and the formation of a weak linear CO band. Although the presence of gaseous NO appears to hinder reductive agglomeration, the rapid removal of oxygen by CO to form CO$_2$ may be the cause for reduction of Rh$^+$ sites to Rh^0 sites that chemisorb linear CO at 2040 cm^{-1}.

The dominance of NO adsorption over CO adsorption on the reduced Rh sites and the inability of NO to adsorb on the oxidized Rh suggest that low wavenumber (1689-1696 cm^{-1}) NO$^-$ on the reduced Rh site may be involved in NO dissociation.

Acknowledgment

GS and RK are grateful for the financial support from the Department of Chemical Engineering, The University of Akron. The authors thank Mr. S. Debnath for assistance in catalyst preparation and collection of part of the IR results.

Literature Cited

1. Taylor, K. C. *Catal. Rev.-Sci. Eng.* **1993**, *35(4)*, 457.
2. Hecker, W. C.; Bell, A. T. *J. Catal.* **1983**, *84*, 200.
3. Oh, S. *J. Catal.* **1991**, *124*, 477.
4. Ng, K. Y. S.; Belton, D. N.; Schmeig, S. J.; Fisher, G. B. *J. Catal.* **1994**, *146*, 394.
5. Yang, A. C.; Garland, C. W. *J. Phys. Chem.* **1957**, *61*, 1504.
6. Yates, J. T. Jr.; Duncan, T. M.; Worley, S. D.; Vaughan, R. W. *J. Chem. Phys.* **1979**, *70*, 1219.
7. Rice, C. A.; Worley, S. D.; Curtis, C. W.; Guin, J. A.; Tarrer, A. R. *J. Chem. Phys.* **1981**, *74*, 6487.
8. Novak, E.; Solymosi, F. *J. Catal.* **1990**, *125*, 112.
9. Arai, H.; Tominaga, H. *J. Catal.* **1976**, *43*, 131.
10. Solymosi, F.; Bansagi, T.; Novak, E. *J. Catal.* **1988**, *112*, 183.
11. Chuang, S. S. C.; Pien, S. I. *J. Catal.* **1992**, *135*, 618.
12. Chuang, S. S. C.; Pien, S. I. *J. Catal.* **1992**, *138*, 536.
13. Srinivas, G.; Chuang, S. S. C. *J. Catal.* (in press).
14. Srinivas, G.; Chuang, S. S., C.; Debnath, S. In *Automotive Emission Catalysis*; Armor, J. N.; Heck, R. M., Eds.; ACS Symposium Series: 1994, Chap. 12; pp 157-167.
15. Srinivas, G.; Chuang, S. S. C.; Balakos, M. W. *AIChE Journal* **1993**, *39*, 530.
16. Debnath, S. M.S. Thesis, The University of Akron, Akron, OH, **1993**.
17. The Standard Sadtler Spectra, Sadtler Research Laboratories, Philadelphia, 1967.
18. Zafiris, G.; Roberts, S. I.; Gorte, R. J.; In *Catalytic Control of Air Pollution*, Silver, R. G.; Sawyer, J. E; Summers, J. C.; Eds.; ACS Symposium Series: 1992, Chap. 6; pp 73-82.
19. Kruse, N; Gaussmann, A. *J. Catal.* **1993**, *144*, 525.
20. Li, Y. E.; Gonzalez, R. D. *J. Phys. Chem.* **1988**, *92*, 1589.
21. Solymosi, F.; Bansagi, T. *J. Phys. Chem.* **1993**, *97*, 10133.

RECEIVED November 16, 1994

Chapter 15

NO Reduction by CO and H_2 over a Pt–CoO$_x$–SiO$_2$ Catalyst

Effect of CoO$_x$ on Activity and Selectivity

Y. J. Mergler, A. van Aalst, and B. E. Nieuwenhuys

Leiden Institute of Chemistry, Gorlaeus Laboratories, Leiden University, P.O. Box 9502, NL-2300 RA Leiden, Netherlands

The reduction of NO by CO and H_2 was studied over a Pt/CoO$_x$/SiO$_2$ catalyst, containing 5 w% Pt and 3 w% Co$_3$O$_4$. For comparison, the same reactions were also studied over 5 w% Pt/SiO$_2$ and 3 w% Co$_3$O$_4$/SiO$_2$ catalysts. The performance of the catalysts was tested using various CO/NO and NO/H_2 ratios, following reductive and oxidative pretreatments. The Pt/SiO$_2$ and Pt/CoO$_x$/SiO$_2$ catalysts showed appreciable activity after a reductive pretreatment at the studied temperature range of 30 - 400°C. CoO$_x$/SiO$_2$ merely formed N_2O after an oxidative pretreatment. The selectivity to N_2 in NO/H_2 reactions was high for the Pt/CoO$_x$/SiO$_2$ catalyst above 225°C, in contrast to the Pt/SiO$_2$ catalyst which showed a high selectivity to NH_3 above 100°C. It is concluded that the addition of CoO$_x$ to Pt results in better catalytic performance in both CO oxidation and NO reduction to N_2.

The reduction of NO_x is one of the most important issues in automotive pollution control. NO_x can be reduced by CO, hydrocarbons or H_2, the latter resulting from the combustion of hydrocarbons and from the water-gas shift reaction (1). Although the reduction of NO by H_2 starts at a lower temperature than by CO, the reaction of CO + NO has also an important contribution to the NO reduction in automotive catalysis because the concentration of CO in the exhaust gas is in general about 10 times higher than the H_2 concentration (1).

The present three-way catalyst (TWC) consists of the noble metals: Pt or Pd and Rh, with Rh being the most efficient catalyst for NO reduction to N_2 (1, 2). At low temperatures, the overall NO conversion over Pt is better than over Rh (3, 4). However, the selectivity of Pt to N_2 formation is poor. In the next years new regulations in the USA and Europe mandate that automotive emission must fall

0097–6156/95/0587–0196$12.00/0

substantially from current levels. The combination of tougher US standards and increased usage of TWC in the world, could potentially strain the Rh supplies. Thus, there is a strong incentive to develop Rh-free automotive catalysts. One possible way may be to combine Pt or Pd with transition metal oxides to achieve better performance of the precious metals in automotive catalysis.

A number of studies have already been directed to the effects of the addition of a metal oxide as a promoter or co-catalyst to Pt or Pd. Kosaki et al. (5) studied the behavior of Pt/Al$_2$O$_3$ catalyst for NO/H$_2$ reactions. Regalbuto et al. (6) and Gandhi et al. (7) used WO$_3$ and MoO$_3$ respectively as an additive to Pt to improve their catalysts for CO/NO reactions. Pd/La$_2$O$_3$/Al$_2$O$_3$ (8), Pd/MoO$_3$/Al$_2$O$_3$ (9) catalysts have also been studied. Recently, Pd-only catalysts have been introduced as commercial TWC (10). This Pd-only catalyst contains Pd as the only noble metal and, in addition, a number of non-classified oxides. It is clear that metal oxides added to the noble metals Pt and Pd may have a strong effect on the catalytic behavior. The success of a Pd-only catalyst (10) supported our opinion that Pt-only catalysts may be developed as a potential substitute for the PtRh TWC.

In our laboratory one research program is directed to studies of the effects of promoters and co-catalysts on the behavior of Pt and Pd catalysts. Other research programs of our group are focused on obtaining a detailed understanding of the specific properties of the current catalyst components. The metal-nitrogen bond strength appeared to be of great importance for the selectivity of the catalyst to N$_2$ formation (4, 11). Rh is more selective to N$_2$ formation than Pt and Pd (2). Based on our earlier results we proposed that under reaction conditions the N concentration is high on a Rh surface whereas it is very low on Pt (11).

In the present paper we describe the performance of Pt/CoO$_x$/SiO$_2$ catalysts. We have made a series of 5 w% Pt/CoO$_x$/SiO$_2$ catalysts, with different cobalt oxide loadings. It was earlier found that the addition of 3 w% Co$_3$O$_4$ resulted in a catalyst that oxidizes CO already at room temperature (12). In the present paper we report results of comparative studies of the CO/NO and NO/H$_2$ reactions over Pt/CoO$_x$/SiO$_2$, Pt/SiO$_2$ and CoO$_x$/SiO$_2$ catalysts.

Experimental

A 5 w% Pt/SiO$_2$ catalysts was made by urea decomposition (13). H$_2$Pt(OH)$_6$ was used as a precursor. After drying, this catalyst was reduced for 3 hours in flowing H$_2$ at 400°C. Part of this reduced catalyst was impregnated with a cobalt nitrate solution resulting, after calcination in air at 400°C, in a catalyst containing 3 w% of Co$_3$O$_4$. (Atomic ratio Pt:Co = 1:1.5). A 3 w% Co$_3$O$_4$/SiO$_2$ catalyst was made for comparison. These catalysts will be referred to as Pt/CoO$_x$/SiO$_2$ and CoO$_x$/SiO$_2$ respectively.

The catalysts were reduced in flowing H$_2$, or oxidized in air, at 400°C for 3 hours, prior to the activity measurements. The measurements were performed in an atmospheric flow apparatus. The gases used were 4 vol% NO/He, 4 vol% CO/He and 4 vol% H$_2$/He (Hoekloos). The flow rate could be adjusted with mass flow controllers to a maximum of 40 ml/min. The CO/NO ratio was varied from oxidizing, CO:NO = 1:2.5, via stoichiometric to N$_2$ with CO:NO = 1:1, to reducing, CO:NO = 3:1. The NO/H$_2$ ratio was 1:1 or 1:3. The temperature was slowly raised

198

Table I. Temperature (°C) required for 25, 50, 75% NO conversion after an oxidative (at 400°C) or reductive (at 400°C) pretreatment and at NO/H$_2$ ratios of 1:1 and 1:3.

Catalyst	Pretreat-ment	Ratio NO:H$_2$	Temp. (°C) 25% conv.	50% conv.	75% conv.
Pt/SiO$_2$	reduction	1:1	60	64	71
Pt/SiO$_2$	reduction	1:3	44	46	50
Pt/SiO$_2$	oxidation	1:1	62	68	80
Pt/SiO$_2$	oxidation	1:3	43	50	52
Pt/CoO$_x$/SiO$_2$	reduction	1:1	37	47	55
Pt/CoO$_x$/SiO$_2$	reduction	1:3	39	45	46
Pt/CoO$_x$/SiO$_2$	oxidation	1:1	97	128	154
Pt/CoO$_x$/SiO$_2$	oxidation	1:3	98	116	125
CoO$_x$/SiO$_2$	reduction	1:1	380	>400	>400
CoO$_x$/SiO$_2$	reduction	1:3	357	397	>400
CoO$_x$/SiO$_2$	oxidation	1:1	287	384	>400
CoO$_x$/SiO$_2$	oxidation	1:3	266	308	>400

with 3°C/min to provide steady state conditions. A quadrupole mass spectrometer was used to monitor the reactant and product gases.

X-Ray diffraction and CO chemisorption were used for catalyst characterization. Both the XRD and CO chemisorption measurements were carried out after a reductive pretreatment. CO chemisorption measurements were performed in a flow apparatus from Quantachrome with H_2 as a carrier gas. CO was injected until the saturation level was reached.

Results and Discussion

a) Catalyst Characterization

Only small Pt particles of an average diameter of 90 Å were found for Pt/SiO_2 and of 70 Å for $Pt/CoO_x/SiO_2$ with XRD. The dispersion of the Pt/SiO_2 catalyst, measured with CO chemisorption, was 10 %, assuming a CO:Pt ratio of 1:1.

b) NO + H_2

The main overall reactions are:

$$2NO + 2H_2 \quad ----> \quad N_2 + H_2O \tag{1}$$
$$2NO + H_2 \quad ----> \quad N_2O + H_2O \tag{2}$$
$$2NO + 5H_2 \quad ----> \quad 2NH_3 + 2H_2O \tag{3}$$

Table I shows the temperatures required for 25, 50, 75 % NO conversion (respectively $T_{25\%}$, $T_{50\%}$, and $T_{75\%}$) after different pretreatments and at NO/H_2 ratios of 1:1 and 1:3. Figures 1 to 3 show the variations in partial pressures of the gases (normalized on the partial pressure of NO at t = 0 s.) with increasing temperature. As can be seen in table I, the addition of CoO_x to a Pt/SiO_2 catalyst had a small beneficial effect on the NO conversion following a reductive pretreatment.

Although the addition of CoO_x appeared to have only a small effect on the conversion of NO, the selectivity to N_2 showed a great improvement above 200°C, when the reaction with a gas ratio of NO:H_2 of 1:3 was considered. Figures 4 to 6 show the selectivity of Pt/SiO_2, $Pt/CoO_x/SiO_2$ and CoO_x/SiO_2 after a reductive and an oxidative pretreatment for NO/H_2 ratios of 1:1 and 1:3.

The NO/H_2 ratio influences the product distribution of the catalysts studied. Differences in the product distribution can be understood on the basis of our earlier results. At low temperatures the NO coverage is high and the rate of dissociation of NO molecules is small (14). The N coverage is low resulting in a low rate of N_2 formation. N_2O formation, which requires a high coverage of molecularly adsorbed NO, is favoured at low temperatures and high NO/H_2 ratios (4, 11). Since ammonia is produced by the addition of hydrogen atoms to N atoms, formed from NO dissociation (4, 11, 15), the production of NH_3 can also be understood on the same grounds. At low NO pressures and excess hydrogen, ammonia can be a major product (4, 11, 15). Lowering the reaction temperature or increasing the NO/H_2 ratio, suppresses the formation of ammonia (4, 11, 15, 16). This is also illustrated by figures 1a, 1b, 2a and 2b. Hardly any NH_3 was formed at a NO/H_2 ratio of 1:1. As could be expected, the selectivity for NH_3 was considerable at a NO/H_2 ratio of

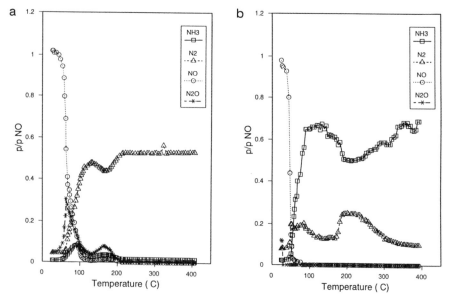

Figure 1.   Formation of N$_2$, NH$_3$ and N$_2$O over Pt/SiO$_2$. a) After a reductive pretreatment, with NO:H$_2$ = 1:1, b) 1:3

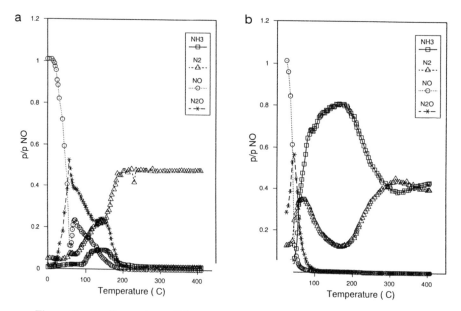

Figure 2.   Formation of N$_2$, NH$_3$ and N$_2$O over Pt/CoO$_x$/SiO$_2$. a) After a reductive pretreatment, with NO:H$_2$ = 1:1, b) 1:3

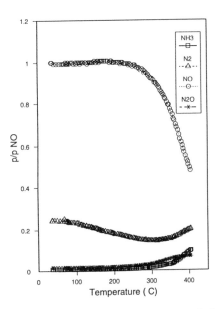

Figure 3. Formation of N_2, NH_3 and N_2O over CoO_x/SiO_2, after a reductive pretreatment, with $NO:H_2 = 1:3$

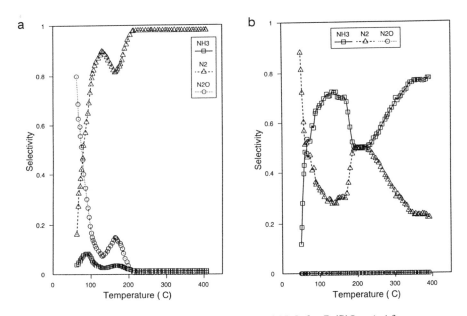

Figure 4. Selectivity to N_2, NH_3 and N_2O for Pt/SiO_2. a) After a reductive pretreatment, with $NO:H_2 = 1:1$, b) 1:3

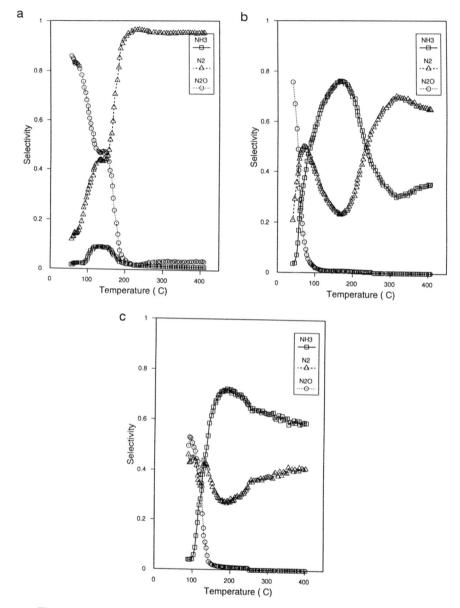

Figure 5.   Selectivity to N$_2$, NH$_3$ and N$_2$O for Pt/CoO$_x$/SiO$_2$. a) After a reductive pretreatment, with NO:H$_2$ = 1:1, b) 1:3, c) after an oxidative pretreatment, with NO:H$_2$ = 1:3

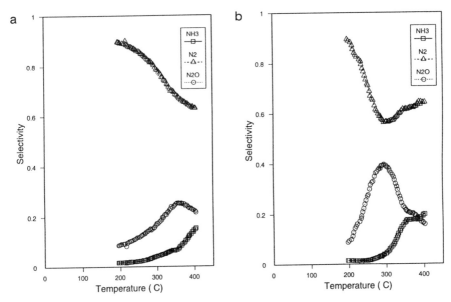

Figure 6.   Selectivity to N_2, NH_3 and N_2O for CoO$_x$/SiO$_2$. a) After a reductive pretreatment, with NO:H_2 = 1:3, b) after an oxidative pretreatment, with NO:H_2 = 1:3

1:3. Upon comparing figure 4b and 5b, it follows that Pt/SiO$_2$ has still a high selectivity to NH$_3$ at 400°C while for Pt/CoO$_x$/SiO$_2$ the selectivity to N$_2$ is higher than to NH$_3$ above 225°C.

CoO$_x$/SiO$_2$ itself is not very active at temperatures lower than 250°C, as can be seen in figure 3. The relatively low T$_{50\%}$ following an oxidative pretreatment is related to the relatively high production of N$_2$O in the temperature range from 250 to 300°C. N$_2$O formation started at lower temperatures over a preoxidized cobalt oxide catalyst, than over a prereduced cobalt oxide catalyst. Figure 6b illustrates that for a pre-oxidized catalyst the selectivity to N$_2$O increases rapidly above 200°C, with a maximum around 300°C. A similar behavior was observed for Pt/CoO$_x$/SiO$_2$. More N$_2$O was produced over Pt/CoO$_x$/SiO$_2$ than over Pt/SiO$_2$ after an oxidative pretreatment.

Meunier et al. studied the CO oxidation reaction and found that Co^{2+} ions in Pt/CoO$_x$/Al$_2$O$_3$ catalysts are more active in oxygen dissociation than Co^{3+} (17). In our case, the addition of CoO$_x$ to Pt appears to provide extra dissociation sites for NO. As can be seen in figure 3, the NO dissociation on CoO$_x$/SiO$_2$ starts around 230°C. The difference in the selectivity to N$_2$ and NH$_3$ with a NO:H$_2$ ratio of 1:3, between Pt/SiO$_2$ and Pt/CoO$_x$/SiO$_2$ is only observed above 225°C. Below that temperature Pt/CoO$_x$/SiO$_2$ behaves in a similar way as Pt/SiO$_2$. Apparently, above 225°C partially reduced CoO$_x$ assists the NO dissociation in such a way that the selectivity to N$_2$ is promoted.

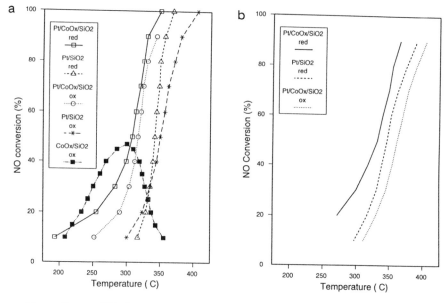

Figure 7a.　NO Conversion over Pt/SiO$_2$, Pt/CoO$_x$/SiO$_2$ and CoO$_x$/SiO$_2$ at a CO:NO ratio of 1:1, following an oxidative or reductive pretreatment

Figure 7b.　NO Conversion over Pt/SiO$_2$ and Pt/CoO$_x$/SiO$_2$ at a CO:NO ratio of 3:1, following an oxidative or reductive pretreatment

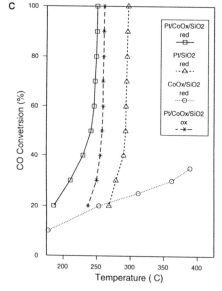

Figure 7c.　CO Conversion over Pt/SiO$_2$, Pt/CoO$_x$/SiO$_2$ and CoO$_x$/SiO$_2$ at a CO:NO ratio of 1:2.5, following an oxidative or reductive pretreatment

As can be seen in figure 5c and in table I, pre-oxidation of the catalyst led to a higher T$_{50\%}$. However, a second series of measurements following oxidation and subsequent cooling to room temperature under the reactant flow, led to lower T$_{50\%}$ (not shown in table I). Apparently, more cobalt oxide was partially reduced during the first series of measurements, and more catalytically active centers were formed (18). Hence, it appears that partially reduced cobalt oxide is more active in NO dissociation and reaction than Co$_3$O$_4$. Tomishige et al., who studied the promoting effect of Sn on NO dissociation and NO reduction with H$_2$ over Rh-Sn/SiO$_2$ catalysts, found that reduced Rh-Sn/SiO$_2$ already dissociates NO at room temperature, but that the rapid dissociation stopped when one third of surface Sn atoms was oxidized by NO. The removal of O atoms on Sn by H$_2$ was the slowest step (19). Since the T$_{50\%}$ of Pt/CoO$_x$/SiO$_2$ after an oxidative pretreatment is much higher than the T$_{50\%}$ of Pt/SiO$_2$, it is likely that the NO dissociation on Pt is retarded by the slow reduction of Co$_3$O$_4$ by H$_2$, just as Tomishige et al. (19) suggested for their Rh-Sn/SiO$_2$ catalyst.

Figure 5c shows that the selectivity to N$_2$ is also worse following oxidation, in comparison to the selectivity to N$_2$ following reduction (fig 5b). Hence, it appears that the available CoO$_x$ centers are less active in NO dissociation after oxidation. Platinum is then the only active species, leading to a higher selectivity to NH$_3$. Cobalt oxide is partially reduced in the reaction mixture at higher temperatures, resulting in an improved selectivity to N$_2$.

Wolf et al. (14) found direct evidence that vacancies on the surface are required for the dissociation of NO molecules. A high coverage of molecularly adsorbed NO inhibits the reaction rate because vacancies are not available. Oh et al. found for Rh catalysts with a high Rh dispersion that the NO dissociation is low, probably because of a low concentration of free neighboring Rh atoms, required for NO dissociation (20).

Cobalt oxide should be partially reduced to provide for the O vacancies required to dissociate NO molecules. Partially reduced active CoO$_x$ centers play an important role to account for the improved selectivity and activity of Pt/CoO$_x$/SiO$_2$ catalyst in NO/H$_2$ reactions. Oxidation of the Pt/CoO$_x$/SiO$_2$ catalyst leads to a higher on-set temperature of the reaction, but the selectivity to N$_2$ increases during the reaction cycle due to the formation of the active CoO$_x$ centers.

c) CO + NO

The overall reactions that can take place are:

$$2CO + 2NO \quad \longrightarrow N_2 + 2CO_2 \qquad (4)$$
$$CO + 2NO \quad \longrightarrow N_2O + CO_2 \qquad (5)$$

Since the masses of CO/N$_2$ and CO$_2$/N$_2$O are equal and the fact that our mass spectrometer is not sensitive enough to monitor masses 14 of N, 16 of O accurately, only the differences in activity of our Pt/SiO$_2$, 3 w% Co$_3$O$_4$/SiO$_2$ and Pt/CoO$_x$/SiO$_2$ catalysts could be measured. The NO conversion (to N$_2$ and N$_2$O) was plotted against the temperature under stoichiometric and reducing conditions while the CO conversion was plotted against the temperature under oxidizing conditions. Figure 7a shows the NO conversion over Pt/SiO$_2$, CoO$_x$/SiO$_2$ and

Pt/CoO$_x$/SiO$_2$ following a reductive and oxidative pretreatment of the catalysts. The CoO$_x$/SiO$_2$ was not active in the temperature range studied after a reductive pretreatment with CO:NO = 1:1. It can be clearly seen that the NO conversion is shifted towards higher temperatures after an oxidative pretreatment. The same conclusion can be drawn upon studying the figures 7b and 7c representing respectively the NO conversion of the prereduced and pre-oxidized catalysts at a CO:NO ratio of 3:1 and CO:NO of 1:2.5.

Figure 7 shows clearly the positive effect of the addition of cobalt oxide to a Pt catalyst on the NO conversion. It was found earlier that when the coverage of NO is high, more N$_2$O is formed than at low NO coverage (4, 11). N$_2$ is formed when the surface of the catalyst is covered with much N$_a$. Hence, at a lower CO/NO ratio, more N$_2$O formation can be expected. Reaction 5 is favored at lower temperatures (4, 11, 21, 22). This explains why the temperatures of NO conversion are lower for CO:NO = 1:2.5. Figure 7a shows a maximum around 300°C for the CoO$_x$ catalyst. This has also been found for the NO/H$_2$ reaction over the same catalyst. The maximum corresponds to the maximum in N$_2$O formation. Hence, most likely, the maximum in conversion observed for the CO/NO reaction around 300°C is also caused by N$_2$O formation. Thus it seems that the main reaction product over the oxidized cobalt oxide catalyst is N$_2$O.

It appears that partially reduced Pt/CoO$_x$/SiO$_2$ catalysts are able to dissociate NO more easily than Pt at a low temperature, leading to a lower on-set temperature for the reactions NO with CO. An oxidative pretreatment lead to a shift to higher temperatures for NO conversion. For a pure CoO$_x$/SiO$_2$ catalyst primarily N$_2$O was formed following an oxidation step.

Conclusions

It was shown earlier that the addition of CoO$_x$ to Pt results in a large increase in CO oxidation activity. The present study shows that the addition of CoO$_x$ to a Pt/SiO$_2$ catalyst also improves the activity in CO/NO and NO/H$_2$ reactions after a reductive pretreatment. In case of an oxidative pretreatment, more N$_2$O is formed over the cobalt oxide containing catalysts. CoO$_x$ enhances the formation of N$_2$O at lower temperatures following an oxidative pretreatment. The selectivity to N$_2$ is greatly enhanced by the addition of CoO$_x$ to Pt/SiO$_2$ for NO/H$_2$ ratio of 1:3. Even after an oxidative pretreatment, the selectivity to N$_2$ increases during the reaction cycle due to the concomitant reduction of the oxide by H$_2$.

References

1. Egelhof, W. F., Jr., in "Chemical Physics of Solid Surfaces and Heterogeneous Catalysis" 4, ch. 9. King, D. A. and Woodruff, D. P., Eds., Elsevier, Amsterdam, **1982**
2. Taylor, K. C., in "Automotive Catalytic Converters", Springer, Berlin **1984**
3. Heezen, L., Kilian, V. N., Van Slooten, R. F., Wolf, R. M. and Nieuwenhuys, B. E., Studies. Surf. Sci. Cat. **1991,** 71, 381

4. Hirano, H., Yamada, T. Tanaka, K. I. Siera, J. and Nieuwenhuys, B. E., in *"New Frontiers in Catalysis"*, Guczi L. et al., Eds., 345, **1993**, (Proc. 10th Int. Congr. Catal., 1992, Budapest)
5. Kosaki, Y., Miyamoto, A. and Murakami, Y., *Bull. Chem. Soc. Jpn.* **1982**, *55*, 1719
6. a) Regalbuto, J. R., Fleisch, T. H. and Wolf, E. E., *J. Catal.* **1987**, *107*, 115 b) Regalbuto, J. R., Allen, C. W. and Wolf, E. E., *J. Catal.* **1987**, *108*, 305 c) Regalbuto, J. R. and Wolf, E. E., *J. Catal.* **1988**, *109*, 13
7. a) Gandhi, H. S., Yao, H. C. and Stepien, H. K., *ASC Symp. Series* **1982**, *178*, 143 b) Gandhi, H. S. and Shelef, M., *Stud. Sur. Sci. Catal.* **1987**, *30*, 199
8. a) Muraki, H., Yokota, K. and Fujitani, Y., *Appl. Catal.* **1989**, *48*, 93 b) Muraki, H., Shinjoh, H. and Fujitani, Y., *Appl. Catal.* **1986**, *22*, 325
9. Halasz, I., Brenner, A. and Shelef, M., *Catal. Lett.* **1992**, *16*, 311
10. Summers, J. C. and Williamson, W. B., in: *Environmental Catal.*, ACS *Symposium Series 552*, Armor, J. N., Ed., **1994**, 94
11. Hirano, H. Yamada, T., Tanaka, K. I., Siera, J., Cobden, P. and Nieuwenhuys, B. E., *Surf. Sci.* **1992**, *262*, 97
12. Mergler, Y. J., Van Aalst, A., Van Delft, J. and Nieuwenhuys, B. E., submitted to: *Stud. Surf. Sci. Catal.: Catalysis and Automotive Pollution Control III*, Proc. 3th Intern. Symp. (CAPoC3), 1994, Frennet, A. and Bastin, J. M., Eds.,
13. Geus, J. W., *Dutch Patent Application 6* **1967**, *705*, 259
14. Wolf, R. M., Bakker, J. W. and Nieuwenhuys, B. E., *Surf. Sci.* **1991**, *246*, 135
15. Otto, K. and Yao, H. C., *J. Catal.* **1980**, *66*, 229
16. Pirug, G. and Bonzel, H. P., *J. Catal.* **1977**, *50*, 64
17. Meunier, G., Garin, F., Schmitt, J., Maire, G. and Roche, R., *Stud. Surf. Sci. and Catal.* **1987**, *30*, 243
18. a) Pande, N. K. and Bell, A. T., *J. Catal.* **1986**, *98*, 7 b) Pande, N. K. and Bell, A. T., *Appl. Catal.* **1986**, *20*, 109
19. Tomishige, K., Asakura, K. and Iwasawa, Y., *J. Chem. Soc., Chem. Commun.* **1993**, 184
20. Oh, S. H., Fisher, G. B., Carpenter, J. E. and Goodman, D. W., *J. Catal.* **1986**, *100*, 360
21. Cho, B. K., Shanks, B. H., and Bailey, J. E., *J. Catal.* **1989**, *115*, 486
22. Kudo, A., Steinberg, M., Bard, A. J., Campion, A., Fox, M. A., Mallouk, T. E., Webber, S. E., and White, J. M., *J. Catal.* **1990**, *125*, 565

RECEIVED November 8, 1994

Chapter 16

The Influence of Activated Carbon Type on NO_x Adsorptive Capacity

A. M. Rubel, M. L. Stewart, and J. M. Stencel

Center for Applied Research, University of Kentucky,
3572 Iron Works Pike, Lexington, KY 40511–8433

The selective capture of NO_x from combustion flue gas by the use of commercially produced activated carbons at typical stack temperatures is reported. This adsorption is independent of whether the NO_x is NO, NO_2 or a mixture. The NO_x adsorption capacities can be as great as 0.15 g NO_x/g carbon if O_2 is present as a co-reactant. NO_2 is the species stored within the carbon and can be released by temperature induced desorption at 140°C. Adsorption capacities are dependent on the type of activated carbon used.

Thermogravimetry/mass spectrometry was used to determine adsorptive capacity of several commercially available activated carbons produced from coal, coconut, and petroleum pitch precursors. The range of their N_2 BET surface areas was between 400 to 2000 m^2/g. Although, carbons with high adsorption capacity contained similar C, N, and O contents, proximate analyses, surface areas and micropore volumes, no significant correlations were found between chemical and physical properties and the NO_x adsorptive capacity. One possibly important characteristic of the carbons correlated with NO_x adsorption capacity was specific and narrow pore size distribution with an effective pore diameter of 0.56 nm.

Our research has shown that activated carbons can be used to selectively capture NO_x (NO and NO_2) from flue gases at typical combustion stack temperatures (70-120°C) (1-3). Temperature programmed desorption releases NO_2 at temperatures near 140°C. It is necessary for O_2 to be present for this selective and large NO_x adsorption capacity, but CO_2 and H_2O do not interfere with adsorption nor are themselves adsorbed to any significant level. The NO_x adsorption capacity can be as high as 0.15 g NO_x/g carbon using a simulated combustion flue gas containing 5% O_2, 15% CO_2, 1% H_2O, 2% NO, balance He

0097–6156/95/0587–0208$12.00/0
© 1995 American Chemical Society

(1-3). At lower NO_x concentrations, the capacities and selectivities are as good, but the rate of adsorption is decreased. Importantly, the same carbon can be recycled through multiple adsorption/ desorption cycles.

In the presence of O_2, adsorption of NO is exothermic, the heat of which is consistent with the conversion of NO to NO_2 at the carbon surface. NO_2 adsorption, with and without the presence of O_2, released less heat than NO adsorption in the presence of O_2 (2,3).

In addition, desorption of the surface adsorbed species produces NO_2 independent of whether NO (with O_2) or NO_2 was the reactant. Hence, and in agreement with other experimentation (4-7), NO_2 is considered to be the adsorbed species and the desorbed product for our conditions of testing. When O_2 was a co-reactant, the adsorption capacity of carbon for either NO or NO_2 was found to be identical. However, without O_2 the adsorption capacity for NO_2 was slightly less than for NO_2 with coreactant O_2, thereby suggesting O_2 also aided the adsorption of NO_2. The temperature of maximum desorption was 10°C higher when NO_2 was used as the adsorbate as compared to NO with coreactant O_2. This difference in desorption temperatures indicates that surface oxygen, adsorbent-adsorbate interactions, or other factors are yet to be defined which determine NO and NO_2 adsorption mechanisms on carbon.

During our earlier work, it was recognized that adsorption capacity of carbons was dependent on the type of activated carbon used. In order to understand the influence of the chemical and physical characteristics of carbons on NO adsorption and to possibly optimize a process for the selective removal of NO_x from combustion flue gas, it was considered important to understand the relationship between adsorption capacity and carbon characteristics, especially the nature of the active binding sites. This work used thermogravimetry/ mass spectrometry (TG/MS) to determine adsorptive capacity of commercially available activated carbons produced from coal, coconut, and petroleum pitch. The relationship between their chemical and physical characteristics relative to NO_x adsorptive capacities was explored.

Experimental

Instrumentation. NO_x adsorption/desorption profiles were obtained using a Seiko TG/DTA 320 coupled to a VG Micromass quadrapole MS. The two instruments were coupled by a heated (170°C) fused silica capillary transfer line leading from above the sample pan in the TG to an inert metrasil molecular leak which interfaced the capillary with the enclosed ion source of the MS. The TG was connected to a disk station which provided for programmable control of the furnace, continuous weight measurements, sweep gas valve switching, data analysis, and export of data to other computers. The MS has a Nier type enclosed ion source, a triple mass filter, and two detectors (a Faraday cup and a secondary emissions multiplier). The MS was controlled by a dedicated personal computer which was also used to acquire and review scans before export to a spreadsheet for data manipulation.

Table I. Physical characteristics of activated carbons studied.

ID	source	SA m²/g	meso SA	micro SA	vol ml/g	meso vol	micro vol
a	coal	460	20	440	0.69	0.45	0.24
b	coco	2000	70	1930	0.89	0.22	0.67
c	coal	850	40	810	0.36	0.05	0.31
d	pitch	685	2	683	0.30	0.04	0.26

Table II. Chemical characteristics of the activated carbons studied.

ID	%mois	%ash	VM	FC	%C	%H	%N	%S	%O (diff)	pH
a	3.1	2.2	17.8	76.8	90.6	0.6	1.3	0.7	6.8	7.4
b	6.0	1.6	48.5	43.9	91.6	0.6	1.4	0.0	6.4	6.2
c	9.8	12.7	19.8	57.7	68.9	1.9	0.8	1.5	26.9	7.0
d	3.2	1.3	22.6	72.9	85.6	1.6	1.6	1.8	9.4	5.3

TG-MS procedures. The TG conditions kept constant during the acquisition of adsorption/desorption profiles were: sweep gas flow rate of 500 ml/min metered at room temperature and pressure, and a constant carbon sample volume weighing approximately 8 to 20 mg. The MS was scanned over a 0-100 amu range with a total measurement interval of approximately 30 s per 100 amu; NO (mass 30) or NO_2 (mass 30 and 46) were identified by comparing amu 30/46 ion ratios.

The details of the TG heating regime used during NO_x adsorption/desorption experiments have been discussed previously (1). Briefly, the regime incorporated segments for outgassing, cooling, adsorption, desorption, and temperature-induced desorption. During outgassing under a He purge, the carbon sample was heated to 400°C, the maximum temperature to be used during desorption. After cooling to the desired adsorption temperatures between 70-120°C, the He flow through the TG was replaced with the simulated flue-gas (see Materials Section). Outgassing and subsequent cooling required approximately 45 to 50 minutes. During adsorption, the carbon was allowed to become saturated, as determined by its mass vs. time profile. The system was then purged of the simulated flue gas. The weight lost during this He purging before the temperature induced desorption step is associated with physically bound adsorbate.

Analytical procedures. Ultimate and proximate analyses were performed on all carbons using LECO instrumentation and standard procedures. Surface area measurements were performed on a Quantachrome Autosorb-6 using a N_2 volumetric flow procedure. Mercury porosimetry measurements were done on a Quantachrome Autoscan porosimeter through a pressure range of 0-60,000 psi. Surface areas were calculated using the standard BET equation between relative pressures of 0.05-0.25 (7). Micropore volumes were calculated using the Dubinin-Radushkavich plot over a linear range (9). Effective micropore size distributions and mean pore diameters were obtained using the Horvath and Kawazoe method (10). The micropore distributions from this method were analyzed within a range of 0.35-1.4 nm. All samples were outgassed under vacuum at approximately 250°C for 16-20 hours prior to analysis.

Materials and simulated flue gas composition. Four commercial carbons produced from a variety of precursors (coal, coconut shells, and petroleum pitch) were used during this study. They were chosen because of their diverse chemical characteristics and wide range of surface areas, 460 to 2000 m^2/g (Tables I and II). All four carbons were produced by steam activation. For the purposes of this paper, the four carbons will be identified by letters **a** to **d** as indicated in Table 1.

The NO_x adsorption capacity of the activated carbons was determined using the following concentrations of gases during adsorption: 2% NO, 5% O_2, 15% CO_2, 0.4% H_2O and He as the balance. Helium was used by itself during degassing, physidesorption, and temperature programmed desorption.

Table III. NO$_x$ adsorption capacity, maximum adsorption and desorption rates

ID	Ads capacity wt% carbon	Max ads rate mmoles NO$_2$/g C/min	Max des rate mmoles NO$_2$/g C/min
a	13.6	0.48	-0.35
b	11.4	0.38	-0.29
c	10.7	0.32	-0.28
d	14.5	0.65	-0.32

Table IV. Coefficients of determination

property	r^2, capacity	r^2, max rate
SA, m^2/g	0.31	0.20
meso SA	0.69	0.65
micro SA	0.29	0.18
vol, ml/g	0.06	0.15
meso vol	0.04	0.01
micro vol	0.29	0.21
%moisture	0.84	0.67
%ash	0.48	0.43
%VM	0.15	0.06
%FC	0.65	0.43
%C	0.27	0.18
%H	0.01	0.01
%N	0.57	0.61
%S	0.10	0.19
%O(diff)	0.36	0.27
pH	0.17	0.42

Results and Discussion

Adsorption capacity, adsorption rates and desorption rates. The NO_x (NO or NO_2) adsorption capacity of the carbons, measured by the weight gain attributed to NO_2, varied from 10.7 to 14.5 weight percent of the carbon (Table III). All carbons became saturated with NO_x in less than 30 minutes. Therefore, the values given in Table III reflect maximum adsorption capacities.

The maximum and initial NO_x adsorption rates for the carbons varied 2-fold and ranged from 0.32 to 0.65 mmoles NO_2/g carbon/min (Table III, Figure 1). Slight differences during the initial response to adsorption in Figure 1 are thought to be related to operation of the two mass flow controllers used to blend the components of the simulated flue-gas.

In contrast to adsorption, the maximum desorption rate during linear temperature programmed desorption (TPD) was nearly the same for all carbons (Table III, Figure 2). The temperature (138°C) of maximum desorption rate was also the same for all carbons (Figure 2). These results suggest that the desorption mechanism is similar for all the carbons and that there is commonality in the adsorption sites or mechanism independent of carbon type. Other recent findings have shown that, at 70°C, 70% of the adsorbed NO_2 can be desorbed (11). Therefore, 70% of the NO_2 formed during adsorption merely condenses or occupies the void space within the pores of the carbon.

Correlations between adsorption capacity and rate vs. the chemical and physical properties of the carbons. A coefficient of determination, r^2, was determined for each characteristic of the carbons which was studied relative to a linear relationship with adsorption capacity or maximum adsorption rate (Table IV). No single characteristic correlated well to adsorption. The best correlation ($r^2=0.84$) was an inverse relationship to moisture content. This possible negative correlation may be related to blockage of NO/NO_2 adsorption sites (11). However, it must be realized that, before NO or NO_2 adsorption, the carbons were pretreated at 400°C under a He purge. Such a temperature would dehydrate the carbon, although oxide and other surface functionality may not be disturbed.

The next highest correlations were an inverse relationship with mesopore surface area and direct relationships with the fixed carbon and nitrogen content of the carbons. However, at a r^2 of between 0.4-0.7, none of these relationships were considered significant. The lack of significant correlation for individual characteristics that were examined does not eliminate the possibility that two or more of these parameters together could control adsorption.

Carbons **a** and **d** had the best adsorption capacities at around 0.14 g NO_2/g carbon. These two carbons had similar C, N, O (by difference) contents, proximate analyses, and comparable total, mesopore, and micropore surface areas, micropore volumes and pore size distributions. All other characteristics determined for carbons **a** and **d** were widely divergent.

The one factor that did emerge as a potentially important characteristic of high capacity carbons was their pore size distribution. Carbons **a** and **d** had

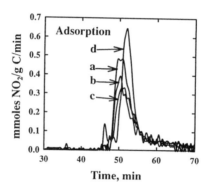

Figure 1. NO$_X$ adsorption rates on activated carbons

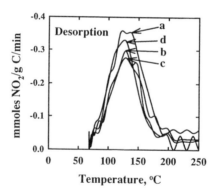

Figure 2. NO$_2$ desorption rates from activated carbons

very similar and narrower pore size distributions with a greater number of pores concentrated around an average effective pore diameter. The mean pore diameters for **a** and **d** were also lower than for the low capacity carbons, **b** and **c** (Figure 3). In contrast, carbons **b** and **c** had comparatively broad pore size distributions. The correlation between adsorption rate or capacity and the average effective pore diameter was better than any other measured property of the carbons (Figure 4). It is also of interest that carbons **b** and **c** had substantially greater surface areas than the other samples. High surface areas are often an important measure of adsorptive capacities of activated carbons, but appear to have little affect on NO$_x$ adsorption capacities.

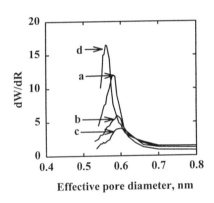

Figure 3. Pore size distribution for activated carbons

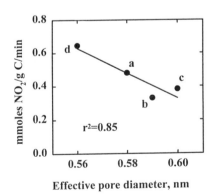

Figure 4. Correlation between NO$_x$ adsorption rate and effective pore diameter of activated carbons

Summary and Conclusions

The results obtained during this study added further to our understanding of NO$_x$ adsorption on activated carbon. Our previous data suggested, that the first step in adsorption was the catalytic conversion by the carbon of NO and O$_2$ to NO$_2$. This step involves an interaction with and/or bonding between the carbon and NO and O$_2$ (1-3). The significant variation measured in the rates of adsorption is compatible with adsorption involving a catalytic step dependent on the nature of the active sites on the carbon. Recent results have shown that 70% of the adsorbed NO$_2$ once formed merely occupies the void space within the pores of the carbon and is not actually chemically bound (11). Therefore, desorption should be primarily diffusion controlled and provided the carbon's pores are above some critical size, desorption rates would be less effected by the chemical characteristics of the carbon consistent with the data.

Since NO adsorption appears to be the two step process of conversion to NO$_2$ and subsequent storage within the pores, the correlation between pore size distribution and adsorption capacity and rate may well be related to the storage capacity and not the active sites.

The understanding of the characteristics of the carbon active binding sites controlling NO$_x$ adsorption will require further investigation. Additional characteristics of the carbon are being considered.

Literature Cited

1. Rubel, A.M.; Stencel, J.M.; Ahmed, S.N. *Preprints Symposium on Flue Gas Cleanup Processes;* ACS, Division of Fuel Chem.: Denver, CO, 1993; 38(2), 726-733.
2. Rubel, A.M.; Stencel, J.M.; Ahmed, S.N. *Proceeding of the 1993 AIChE Summer National Meeting*; AIChE: Seattle, WA, 1993; Paper no. 77b.
3. Rubel, A.M.; Stewart, M.L.; Stencel, J.M. *J.M.R.*, submitted July, 1994.
4. Richter, E.; Schmidt, H.J.; Schecker, H.G. *Chem. Eng. Tech.* 1990, *13*, 332.
5. Kaneko, K.; Camara, S.; Ozeki, S.; Souma, M. *Carbon* 1991, *29*, 1287.
6. Kakuta, N.; Sumiya, S.; Yoshida, K. *Catal. Letters* 1991, *11*, 71.
7. Rao, M.N.; Hougen, O.A. *Chem. Eng. Prog.* 1952, *48*, 110.
8. Brunauer, S.; Emmett, P.H.; Teller, E. *J.A.M. Chem. Soc.* 1938, *60*, 309.
9. Dubinin, M.M.; Radushkavich, L.V. *Proc. Acad. Sci.* (USSR), 1947, *55*, 331.
10. Horvath, G.; Kawazoe, K.J. *J. Chem. Eng. Japan*, 1983, *16*(6), 470.
11. Rubel, A.M.; Stencel, J.M. University of Kentucky, Center for Applied Energy Research, unpublished data.

RECEIVED September 19, 1994

Chapter 17

Applications of Urea-Based Selective Noncatalytic Reduction in Hydrocarbon Processing Industry

M. Linda Lin, Joseph R. Comparato, and William H. Sun

Nalco Fuel Tech, P.O. Box 3031, Naperville, IL 60566–3031

Urea-based SNCR process has been applied successfully on refinery CO boilers, process heaters, package boilers, and waste incinerators. Moderate to high NO_x reductions (50-95%) with minimum NH_3 slip $\leq$ 15 ppm were achieved under wide temperature and load conditions. Good performance was demonstrated when the combustion sources are characterized by low and high NO_x baselines generated by regular or waste fuels, and under large load fluctuations. Comprehensive process design and engineering make highest utilization of the chemical reagent possible. The urea-based SNCR technology has expanded the options that refinery and petrochemical plants have in NO_x control strategies.

Combustion in the hydrocarbon processing industry produces a wide range of nitrogen oxides (NO_x) emissions from a large variety of sources. The wide range of NO_x emissions in a petrochemical plant and refinery could come from sources such as CO boilers, ethylene crackers, process heaters, steam boilers, turbines, as well as waste incinerators. The hydrocarbon processing industry is challenged by the task of developing a NO_x emission control strategy which considers both high and low emission sources, and at the same time reduces the emission compliance cost while meeting current and future NO_x regulations. Recent advances in selective non-catalytic reduction (SNCR) technology have found success in many full-scale combustion systems burning solid, liquid and gaseous fuels. Consequently, many states have recognized urea-based SNCR as a flexible and cost-effective NO_x reduction process.

0097–6156/95/0587–0217$12.00/0

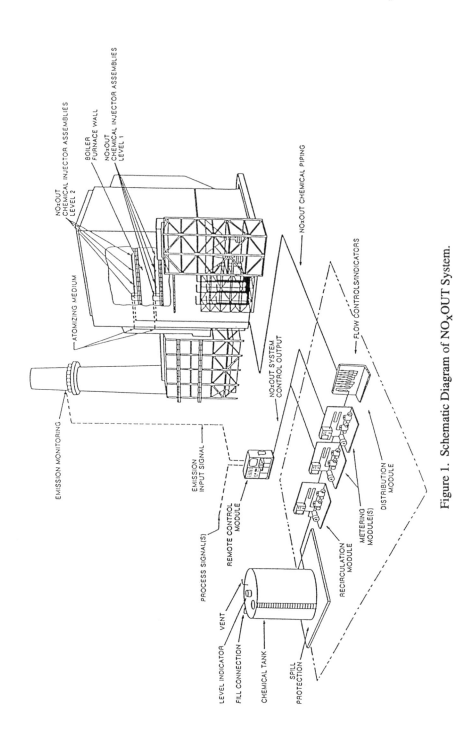

Figure 1. Schematic Diagram of NO$_X$OUT System.

The Urea-Based SNCR Technology

A Major NO_x Abatement Technology.   Post-combustion NO_x treatment, particularly the NO_xOUT process (Figure 1) *(1)*, is a representative urea-based SNCR technology.   The technology initially emerged from research on the use of urea to reduce nitrogen oxides by the Electric Power Research Institute (EPRI). Currently, there are thirty three (33) U.S. patents centered around the NOxOUT process, and thirty one (31) of them were granted after 1988.   The urea-based SNCR has joined other NO_x abatement methods such as combustion modifications and selective catalytic reduction (SCR), as a major cost-effective NO_x abatement technology.   The NO_xOUT process has been applied to over one hundred stationary systems with good NO_x reduction and relatively low cost.   Due to continuous mechanical and chemical improvements, the process now can achieve moderate to high NO_x reductions (50-70% typical, and can be as high as 80-90%) with minimal secondary by-products ($\leq$ 15 ppm NH_3).

Comparison with Use of Ammonia.   Unlike ammonia, urea-based SNCR utilizes an innocuous chemical, relieving safety and handling concerns of gaseous and aqueous ammonia.   Like NH_3 used in SNCR and SCR, urea has a discrete temperature window for the process to be effective.   The urea-SNCR temperature window is typically between 871 and 1,121 °C or 1,600 and 2,050 °F (Figure 2), which can be found in the furnace between burners and economizer.   The overall chemical reaction for urea to reduce nitric oxide is expressed as:

$$H_2NCONH_2 + 2\ NO + 1/2\ O_2 \rightarrow 2\ N_2 + CO_2 + 2\ H_2O\ .$$

As one mole of urea reacts with two moles of NO, a normalized stoichiometric ratio (NSR) of 1 is defined.   The temperature window for NH_3 SNCR is almost identical to that of urea, but the NH_3 SNCR window shifts slightly lower (ca. 55 °C or 100 °F).   The urea-based SNCR process injects an aqueous urea solution, which provides an advantage over ammonia.   Using liquid droplets, the release of urea can be controlled to match temperature and NO_x gradient in a furnace.   The evaporation rate and penetration of urea droplets are designed with proper particle size and momentum which fit furnace configurations on a case by case basis.   This capability improves chemical reagent utilization and eliminates use of mechanical mixing devices such as the ammonia injection grid (AIG) system or large quantities of a carrier gas such as air, steam or flue gas recycle.   When inappropriately applied, injected urea, as well as ammonia, can be oxidized at excessively high temperatures (e.g. > 1,204 °C or 2200 °F).   This will result in an increase in NO_x formation rather than NO_x destruction.   At low temperatures, by-product NH_3 levels can become significant.   This is called NH_3 slip (Figure 3).

Critical Process Parameters.   In addition to temperature, residence time is also a critical process parameter for SNCR.   Longer residence time allows a higher degree

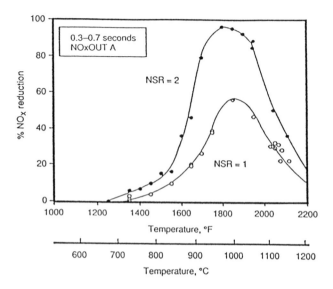

Figure 2. Influence of Temperature on NO$_x$ Reduction in Urea-Based SNCR Process.

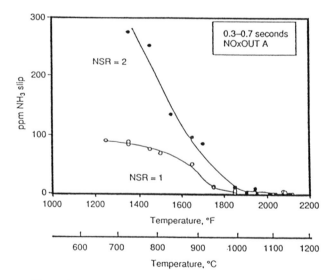

Figure 3. Ammonia Slip as a Function of Temperature.

of completion of urea reaction with NO_x. Longer residence time also favorably broadens the application temperature window (Figure 4). Applications in refinery furnace heaters and municipal waste combustors are easier applications for SNCR because of their generally longer residence time. The urea and NO_x flame reaction mechanism is very complex. Major reaction pathways can be categorically classified into an ammonia and a cyanuric acid route *(2-5)*. The ammonia route produces ammidozine ($NH_2\cdot$) radicals which then react with NO to form the desirable product nitrogen. The cyanuric acid route leads to intermediates, such as isocyanic and cyanic acid, which can lead to a small amount of secondary byproducts such as nitrous oxide (N_2O) and carbon monoxide (CO).

Carbon monoxide concentration at the point of chemical injection is also important *(2)*. Carbon monoxide is a measure of the gas phase hydroxyl radical (OH·) concentration, an important intermediate in the NOx reduction chemistry. Increasing OH· radical concentration has the effect of shifting the temperature window to a lower temperature for either urea or ammonia based systems. In some cases this can be a disadvantage but in many cases this knowledge can be used for further optimization of the NO_x reduction process.

Key Process Features. The NO_xOUT program incorporates various scientific methods to accommodate non-ideal situations relative to temperature and residence time. First, influence of common flue gas components on NO_x reduction as a function of temperature and residence time is experimentally and theoretically determined. These include combustion simulations using pilot combustors and computational fluid dynamic and chemical kinetic modeling. Next, chemical injections at multiple locations with automatic process control allows more efficient response to temperature and load changes. Then, special injector and nozzle designs are used to meet liquid droplet size and chemical distribution requirements for the specific application *(6)*. The urea-based chemical formulations insure consistent product quality control and include specialty additives which prevent problems such as injector fouling. With these capabilities, urea spray trajectories can closely follow temperature profile in the post combustion zone, resulting in a high chemical reaction efficiency. In practice, the accurate predictability minimized field optimization time. With pre-assembled feed equipment comprised of modules, the system offers ease of installation and maximal operational flexibility. Overall, the process performance can typically be demonstrated in two weeks, and installation finished within a time period that does not interfere with normal operation.

The byproducts that have been observed for SNCR are ammonia, carbon monoxide, and nitrous oxide. All three byproducts are minimized when the temperature is on the high side of the window and residence time is relatively long. Ammonia is a concern because of the possibility of forming ammonium salts (sulfate, bisulfate, and chloride). These salts can foul backend heat exchange equipment or form visible

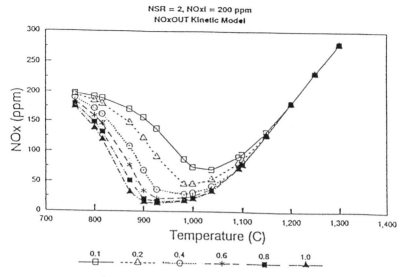

Figure 4. Influence of Residence Time on NO$_x$ Reduction in Urea-Based SNCR Process.

plumes. Carbon monoxide is normally sufficiently low (< 20 ppm) so that it does not create a problem. Nitrous oxide is not included in any regulatory definition of NO$_x$ but is of concern as a potential greenhouse gas. Formulation of N$_2$O can range from zero to 15% of the NO$_x$ reduced. Typically, byproducts have been maintained so as to not create any operational or environmental concerns.

New Advances in Urea-Based SNCR. More recently, several advanced urea-based technologies were developed to add greater versatility to the NO$_x$ compliance strategy employing SNCR. Exemplary advanced processes include the SNCR/SCR Hybrid (7), the NO$_x$OUT PLUS process (8), and a urea-slurry process consisting of injecting urea-based chemical with an alkaline slurry (9). The urea and alkaline slurry process was designed to simultaneously remove NO$_x$, SO$_x$, and HCl by injecting slurry into the furnace. The NO$_x$OUT PLUS process involves an in-situ thermal treatment of urea, which produces a more kinetically active reagent. The enhanced reagent widens the temperature window, in combination with or absence of urea. This treatment also has the advantage of reducing emissions of nitrous oxide and carbon monoxide compared to urea. The SNCR/SCR Hybrid process was designed to achieve high NO$_x$ reduction efficiency similar to SCR without the hazards associated with ammonia and with a lower cost than SCR. The pilot test results demonstrated that the ammonia generated from the SNCR process is an effective reducing agent for a subsequent down-sized catalyst. Since SNCR is used to reduce the majority of NO$_x$, catalyst requirements are reduced. Space velocity and pressure drop lead to a smaller catalyst size which can normally be fitted or designed into a duct between an economizer and air preheaters.

Urea-Based SNCR Applications

The urea-based SNCR process can be used to meet the requirements for NO_x abatement in the hydrocarbon processing industry.   The combustion units in refineries and petrochemical plants are usually operated within the effective SNCR temperature window, and with residence time meeting or exceeding the minimum required (Table I).   The function of a CO boiler in a refinery is to burnout the FCC regenerator offgas containing significant concentrations of CO and some volatile hydrocarbons.   Process heaters perform heat exchange functions, normally at relatively low heat intensities.   Waste incinerators are operated to completely burnout organic waste and hazardous materials.   Steam boilers in refineries and petrochemical plants produce steam for use in various processes and power generation.   The steam boiler aspect of urea-based SNCR application is the same as those previously described *(10-12).*

Table I. Parameters Influencing Urea-Based SNCR Applications in Refineries

	CO Boiler	Process Heaters	Waste Incinerators
Temperature, °C	982-1093	871-1038	927-1093
(Temperature, °F)	1800-2000	1600-1900	1700-2000)
Residence Time, sec	0.3-0.6	0.4-0.8	0.3-1.5
Flue Gas Velocity, m/sec	4.6-7.6	0.6-2.4	6.1-9.1

CO Boiler. This type of boiler burns CO-laden regenerator offgas that exits a Fluidized Catalytic Cracking (FCC) unit.   Depending upon the steam demand, supplementary firing with refinery off-gas is instituted.   Consequently, these units experience a large fluctuation in load swings under normal operating conditions. As burnout within the FCC regenerator has improved in recent years, the CO content of the fuel has declined in many cases. The heat input from CO has been replaced by a variety of refinery fuel gas streams.

Urea-based SNCR performance for a CO boiler has been characterized at high 136,000 Kg/hr or 300,000 lb/hr), medium, and low load, plus "normal" (95,200 Kg/hr or 210,000 lb/hr) conditions. The five observation ports on the CO boiler (Figure 5) were utilized as injection ports. The NO_x baseline under high load condition was approximately 130 ppm.   This was reduced to 60 ppm with urea SNCR which represents a 54% NO_x reduction. NO_x reductions were in the range of 45-55% at other operating conditions (Figure 6). This far exceeds the minimum NO_x reduction requirement of 24%. Ammonia slip was less than 20 ppm, as measured at the ESP inlet, and was 2-6 ppm at lower loads. The urea-based SNCR performance demonstrated its capability of   following frequent load and temperature changes in an adverse low NO_x baseline situation (Figure 6), (one lb/MMBtu equals 1/2326.3 Kg/MJ).

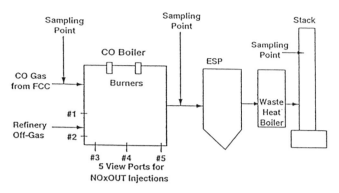

Figure 5. Urea-Based SNCR Application in CO Boiler.

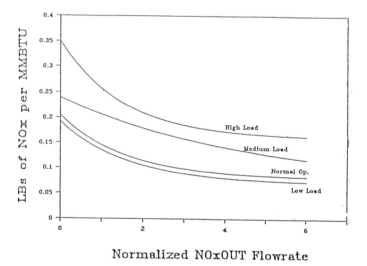

Figure 6. NO$_x$ Reduction at Various Loads in CO Boiler.

An oxygenated hydrocarbon enhancer was also tested by co-injecting it with the urea-based chemical. The co-injection resulted in 15% additional NO$_x$ reduction at lower temperatures (816-927 °C or 1,500 - 1,700 °F). Particulate measurement at the ESP showed that the urea-based SNCR had no effect on total particulate load or its collection efficiency.

Process Heaters. The subject process heaters are located in a refinery which is in an ozone non-attainment area. The No. 1 process heater burns refinery off-gas at 52,750 MJ/hr or 50 MMBtu/hr heat input. The No. 2 heater fires at 186,735 MJ/hr or 177 MMBtu/hr with refinery off-gas and supplemental natural gas. The use of urea-based SNCR reduced NO$_x$ emissions of No. 1 furnace heater from approximately 100 ppm to 30 ppm or 70% (Figure 7). The NO$_x$ emissions of No. 2 furnace heater was reduced from 90 ppm to 38 ppm or 58% (Figure 8). This

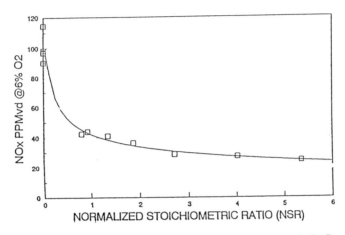

Figure 7. NO$_x$ Reduction as a Function of Reagent Dosage in Refinery Furnace Heater No.1.

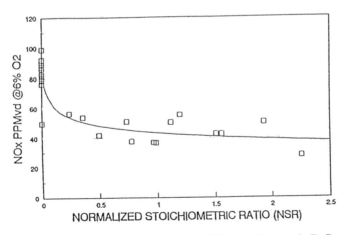

Figure 8. NO$_x$ Reduction as a Function of Reagent Dosage in Refinery Furnace Heater No.2.

demonstrated the capability of urea-based SNCR in reducing NO$_x$ from   low baselines of NO$_x$ in furnace heaters.

The actual tons per year of NO$_x$ reduced from the No. 2 heater were actually more than the No. 1 heater because of the larger size of No. 2. This is an important factor in the NO$_x$ control strategy dealing with individual and total emissions. For this application, the NO$_x$ reductions were obtained in order to provide offsets for expansion of refinery capacity.

Waste Incinerator. Some refinery processes produce a waste sludge containing high levels of acetonitrile (ACN), acrylonitrile (AN), and other waste materials. In the U.S., the ACN and AN in the water constitute a hazardous waste that requires incineration to destruct 99.99% of the ACN and AN.

The urea-based SNCR process was applied on two similar systems consisting of an absorber offgas (AOG) incinerator and a HRSG (Heat Recovery Steam Generator) burning AOG and an aqueous waste stream containing ACN and AN. A diagram of the AOG incinerator and the injector locations of the urea-based chemical is shown in Figure 9. As noted in the Figure, the AOG ports surround the burner, and so does the Overfire Air (OFA).   Through an injector, aqueous ACN/AN stream is atomized to the tip of the burner, where primary combustion air and fuel gas are also added. The AOG fume characteristics are listed in Table II. Combustion of fuel gas incinerates the ACN/AN stream as well as the AOG. At the end, the cylinder narrows as gas exits to the HRSG.   To insure complete incineration, the water walls of the waste heat boiler are covered with refractory. The incinerator firing rate is controlled by maintaining the temperature at the exit of the incinerator to a specified level. The baseline NO$_x$ varied between 13.6 to 81.6 Kg/hr (30 to 180 lb/hr) or 10 to 350 ppm, depending on the firing rate and ACN/AN flow. The plant intended to control the NO$_x$ level to a limit of 49.4 Kg/hr or 109 lb/hr.

Table II. Absorber Off Gas Fume Characteristics

Component	Normal Volume %	Range Volume %
N$_2$	94.17	92.0-95.0
CO$_2$	2.06	1.5-2.5
H$_2$O	1.76	1.5-2.0
CO	1.24	1.0-1.5
Propane (C$_3$H$_8$)	0.62	0.5-1.2
Propylene (C$_3$H$_6$)	0.14	0-2.0
O$_2$	0.02	0-2.0
HCN ppm$_v$	65	40-200
Acrylonitrile ppm$_v$	10	5-150
Acetonitrile ppm$_v$	5	2-100

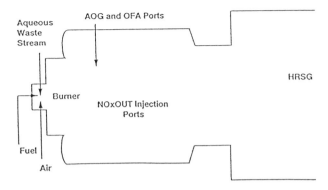

Figure 9. Diagram of Urea-Based SNCR Application in Absorber Off Gas-Heat Recovery Steam Generator Incinerator.

Baseline NO_x increased as temperature increased. They were 132, 213, and 261 ppm at 735, 871, and 982 °C (1,355, 1,600, 1,800 °F), respectively. The NO_x reduction results were better at the higher temperatures. When the ACN/AN waste stream was incinerated, the urea-based SNCR provided up to 80% NO_x reduction at the NSR of 1.9 and 982 °C or 1,800 °F (Table III). When the feed of waste stream was off, NO_x reduction was as high as 83% at 732 °C or 1350 °F and the NSR of 2, and 97% at 871 °C or 1600 °F and at the NSR of 1.9 (Table IV). Ammonia slip was less than 5 ppm. The process reduced NO_x consistently to levels below the emission limit of 49.4 Kg/hr (109 lb/hr). In fact, the NO_x levels were from 9.1 to 68.0 Kg/hr (20 to 150 lb/hr) on Unit #1, and from 22.7 to 72.6 Kg/hr (50 to 160 lb/hr) on Unit #2. This case demonstrated the high performance of urea-based SNCR in a refinery incinerator.

Table III. NOxOUT Performance in Absorber Off Gas Incinerator with Acetonitrile and Acrylonitrile Feed

Temperature = 982 °C (1,800 °F)

NSR	ppm NO_x	% NO_x Reduction
0	261	0
0.5	180	31
1	140	46
1.5	86	67
1.9	53	80

**Table IV. NOxOUT Performance in Absorber Off Gas Incinerator
without Acetonitrile/Acrylonitrile Feed**

I. Temperature = 732 °C (1,350 °F)

NSR	ppm NO$_x$	% NO$_x$ Reduction
0	334	0
0.5	236	29
1	180	46
1.4	109	66
2	57	83

II. Temperature = 871 °C (1,600 °F)

NSR	ppm NO$_x$	% NO$_x$ Reduction
0	327	0
0.5	215	34
1	125	62
1.5	36	89
1.9	10	97

Compliance Strategy

In the hydrocarbon processing industry, the urea-based SNCR technology has shown to be capable of controlling NO$_x$ emissions from high and low NO$_x$ baselines, under large load fluctuations, as well as with a variety of fuels. Furthermore, the urea-based SNCR can work together with other major NO$_x$ control methods such as combustion modifications and SCR. These capabilities offer a compliance strategy which can satisfy the current and future NO$_x$ regulations. The major NO$_x$ emissions from existing sources are concentrated on CO boilers, ethylene crackers, and some heavy oil heaters. Reducing NO$_x$ emissions from these units at high levels may be sufficient enough for the entire plant under the bubble approach. The capability of over-control also may be used as an offset to allow new units to be permitted. In southern California, and perhaps soon in Texas and in other environmentally enlightened states, extra NO$_x$ tons reduced can be used for future emission reduction credits.

Conclusions

The urea-based NO$_x$OUT process has demonstrated high NO$_x$ reduction while minimizing NH$_3$ slip in CO boilers, process heaters, and waste incinerators located in refineries and petrochemical plants. The high performance was achieved under wide temperature and load conditions, as well as low and high NO_x baselines generated by regular and waste fuels. The simplicity, flexibility, and versatility of the process allow formation of a compliance strategy that reduces cost while satisfying the current and future NO_x regulations.

Literature Cited

1. NO_xOUT is a registered trademark of Fuel Tech, Inc. Refer to "The NO_xOUT Process," NFT-2, publication by Nalco Fuel Tech.
2. Sun, W. H.; Hofmann, J. E. "Reaction Kinetics of Post Combustion NO_x Reduction with Urea," *Amer. Flame Research Committee*, 1991 Spring Meeting, Hartford, CT, March, 1991.
3. Michels, W. F.; Gnaedig, G.; Comparato, J. R. "The Application of Computational Fluid Dynamics in the NO_xOUT Process for Reducing NO_x Emissions from Stationary Combustion Sources," *Proceedings of Amer. Flame Research Committee International Symposium*, San Francisco, CA, October, 1990.
4. Teixeira, D. P.; Muzio, L. J.; Montgomery, T. A.; Quartucy, G. C.; Martz, T. D. "Widening the Urea Temperature Window," 1991 Joint EPA/EPRI Symposium on Stationary Combustion NO_x Control, Washington, D. C., March, 1991.
5. Muzio, L. J.; Montgomery, T. A.; Quartucy, G. C.; Cole, J. A.; Kramlich, J. C. "N_2O Formation in Selective Non-Catalytic NO_x Reduction Processes," 1991 Joint EPA/EPRI Symposium on Stationary Combustion NO_x Control, Washington, D. C., March, 1991.
6. Diep, D. V.; Michels, W. F.; Lin, M. L. "Enhanced Spray Atomization in the NO_xOUT Process," *85th Annual Air & Waste Management Association Meeting*, Kansas City, MO, June, 1992.
7. Lin, M. L.; Diep, D. V. "The SNCR/SCR Hybrid NO_x Emissions Control Technology," *Proceedings of the Tenth Pittsburgh Coal Conference*, Pittsburgh, PA, September, 1993.
8. Lin, M. L.; Diep, D. V. "NO_xOUT PLUS - An Improved SNCR Process for NO_x Emission Control," *AIChE Annual Meeting*, Miami Beach, FL, Nov., 1992.
9. Lin, M. L., "New Advancements in NO_x Control Technologies," *Proceedings of CAAPCON*, Chicago, Illinois, July 2-5, 1993.
10. Hofmann, J. E.; von Bergmann, J.; Bokenbrink, D.; Hein, K. "NO_x Control in a Brown Coal-Fired Utility Boiler," *EPRI/EPA Symposium on Stationary Combustion NO_x Control*, (NTIS PB89-220537), March, 1989 (June, 1989).
11. Comparato, J. R.; Buchs, R. A.; Arnold D. S.; Bailey, L. K. "NO_x Reduction at the Argus Plant Using the NO_xOUT Process," 1991 Joint EPA/EPRI Symposium on Stationary Combustion NO_x Control, Washington, D. C., March, 1991.
12. Lin, M. L.; Comparato, J. R.; Lo, C. S. "Use of Non-Catalytic NO_x Treatment in Asia," *Power-Gen Asia*, September, 1993, Singapore.

RECEIVED October 31, 1994

Author Index

Affiliation Index

Subject Index

Bestsellers from ACS Books

The ACS Style Guide: A Manual for Authors and Editors
Edited by Janet S. Dodd
264 pp; clothbound ISBN 0–8412–0917–0; paperback ISBN 0–8412–0943–X

Understanding Chemical Patents: A Guide for the Inventor
By John T. Maynard and Howard M. Peters
184 pp; clothbound ISBN 0–8412–1997–4; paperback ISBN 0–8412–1998–2

Chemical Activities (student and teacher editions)
By Christie L. Borgford and Lee R. Summerlin
330 pp; spiralbound ISBN 0–8412–1417–4; teacher ed. ISBN 0–8412–1416–6

Chemical Demonstrations: A Sourcebook for Teachers,
Volumes 1 and 2, Second Edition
Volume 1 by Lee R. Summerlin and James L. Ealy, Jr.;
Vol. 1, 198 pp; spiralbound ISBN 0–8412–1481–6;
Volume 2 by Lee R. Summerlin, Christie L. Borgford, and Julie B. Ealy
Vol. 2, 234 pp; spiralbound ISBN 0–8412–1535–9

Chemistry and Crime: From Sherlock Holmes to Today's Courtroom
Edited by Samuel M. Gerber
135 pp; clothbound ISBN 0–8412–0784–4; paperback ISBN 0–8412–0785–2

Writing the Laboratory Notebook
By Howard M. Kanare
145 pp; clothbound ISBN 0–8412–0906–5; paperback ISBN 0–8412–0933–2

Developing a Chemical Hygiene Plan
By Jay A. Young, Warren K. Kingsley, and George H. Wahl, Jr.
paperback ISBN 0–8412–1876–5

Introduction to Microwave Sample Preparation: Theory and Practice
Edited by H. M. Kingston and Lois B. Jassie
263 pp; clothbound ISBN 0–8412–1450–6

Principles of Environmental Sampling
Edited by Lawrence H. Keith
ACS Professional Reference Book; 458 pp;
clothbound ISBN 0–8412–1173–6; paperback ISBN 0–8412–1437–9

Biotechnology and Materials Science: Chemistry for the Future
Edited by Mary L. Good (Jacqueline K. Barton, Associate Editor)
135 pp; clothbound ISBN 0–8412–1472–7; paperback ISBN 0–8412–1473–5

For further information and a free catalog of ACS books, contact:
American Chemical Society
Product Services Office
1155 16th Street, NW, Washington, DC 20036
Telephone 800–227–5558